Traeger Grill & Smoker Cookbook For Beginners And Experts

Cook meat like a pro, begin to wow your friends and family with amazing and mouthwatering recipes

Adam Green

Copyright 2021 - All rights reserved.

The content contained within this book may not be reproduced, duplicated or transmitted without direct written permission from the author or the publisher.
Under no circumstances will any blame or legal responsibility be held against the publisher, or author, for any damages, reparation, or monetary loss due to the information contained within this book. Either directly or indirectly.

Legal Notice:
This book is copyright protected. This book is only for personal use. You cannot amend, distribute, sell, use, quote or paraphrase any part, or the content within this book, without the consent of the author or publisher.

Disclaimer Notice:
Please note the information contained within this document is for educational and entertainment purposes only. All effort has been executed to present accurate, up to date, and reliable, complete information. No warranties of any kind are declared or implied. Readers acknowledge that the author is not engaging in the rendering of legal, financial, medical or professional advice. The content within this book has been derived from various sources. Please consult a licensed professional before attempting any techniques outlined in this book.

By reading this document, the reader agrees that under no circumstances is the author responsible for any losses, direct or indirect, which are incurred as a result of the use of information contained within this document, including, but not limited to, errors, omissions, or inaccuracies.

Table Of Contents

Introduction 8
Smoked Beef Recipes 10
1. Porterhouse Steaks with creamed Greens 10
2. Buttered smoked Porterhouse Steak 12
3. Grilled Bloody Mary Flank Steak 14
4. Smoked Tomahawk Steak 16
Smoke Pork Recipes 18
5. Light and Lean Loin 18
6. Beer-smoked Brats 21
7. Carolina Hash 23
8. Polynesian Sauce-stuffed Pork Chops 25
9. Chinese-Style sticky red Ribs 27
Smoke Lamb Recipes 30
10. Pistachio crusted Lamb with Vegetables 30
11. Smoked Lamb Leg with Salsa Verde 33
12. Armenian Style Lamb Shanks with Barley Risotto 36
13. Juicy Mint Pulled smoked Lamb Shoulder 39
Smoked Poultry Recipes 42
14. Smoked Chicken Tikka Drumsticks 42
15. Peruvian Roasted Chicken with Green Sauce 45
16. Roasted Tingle Wings 48
17. Grilled Asian Lettuce Wrap Turkey Burger 51
18. Roasted Tinfoil Dinners 53
19. BBQ Chicken baked Potato 55
20. Grill-fried Buttermilk Chicken with Peppercorn Gravy 57
Extra Smoke Recipes 60
21. Wild Boar 60

22.	Honey Apricot Smoked Lamb Shank	62
23.	Braised Rabbit and Red Wine Stew	64
24.	Citrus Smoked Goose Breast	66
25.	Maple-Glazed Pheasants	68
26.	Ultimate Duck Breasts	70

Smoked Fish and Seafood Recipes 72

27.	Mussels with Pancetta Aioli	72
28.	Mango Shrimp	74
29.	Roasted wild Salmon with Pickled Cauliflower Salad	76
30.	Blackened Catfish	79
31.	Juicy Smoked Salmon	81
32.	Peppercorn Tuna Steaks	83
33.	Stuffed Shrimp Tilapia	85
34.	Grilled Shrimp	88

Smoked Vegetable Recipes 90

35.	Baked Breakfast Mini Quiches	90
36.	Baked Creamed Spinach	93
37.	Smoked Baked Potato Soup	96
38.	Smoked Whole Pickles	98
39.	Grilled Sweet Potato Planks	100
40.	Roasted Veggies & Hummus	102

Conclusion 104

Introduction

Thank you very much for purchasing this cookbook. If you want a unique barbecue experience, check out these recipes. The dishes that I will offer you will be really tasty and I am sure that with my advice you will have fun recreating them. I hope you will find satisfaction in preparing the many dishes in this book, when you do not know what to cook look at my recipes, you will immediately find inspiration and I am sure that you will become a real chef.

Enjoy your meal.

Smoked Beef Recipes

1. **Porterhouse Steaks with creamed Greens**

Preparation time: 10 minutes

Cooking time: 1 hour and 20 minutes

Servings: 4

Ingredients

- 2 porterhouse steaks
- Kosher salt and cracked black pepper, to taste
- 6 tablespoons butter, divided
- 1 shallot, thinly sliced
- 2 clove garlic, minced
- 1 cup heavy cream
- 1 pinch ground nutmeg

- 3 pounds (1.4 kg) mixed salad greens

Direction

1. Preheat the grill smoker to 300 °F (150 °C).

2. Pat steaks dry and season with salt and pepper to taste. Set aside on a baking sheet. Place in the oven until perfectly medium, about 1 hour and 20 minutes. Remove from the oven and let rest on a cutting board for about 10 minutes. Cut steaks into 3/4-inch (1.9 cm) pieces.

3. Add remaining butter and shallots to a hot pan. Add minced garlic and bring to medium heat, stirring constantly. Let the garlic release its aroma, about 20 seconds.

4. Add cream, simmer, stirring frequently, and ground nutmeg. Turn off heat and set aside.

5. In a large mixing bowl, toss greens with some dressing or vinaigrette. Add steak pieces as well as some of the shallots and garlic. Serve!

Nutrition:
- **Energy (calories):** 2306 kcal
- **Protein:** 62.48 g
- **Fat:** 215.31 g
- **Carbohydrates:** 27.72 g
- **Calcium:** 197 mg
- **Magnesium:** 57 mg
- **Phosphorus:** 644 mg
- **Iron:** 6.53 mg

2. Buttered smoked Porterhouse Steak

Preparation time: 15 minutes

Cooking time: 45 minutes

Servings: 2

Ingredients:

- 4 tablespoons butter, melted
- 2 teaspoons Dijon mustard
- 2 tablespoons Worcestershire sauce
- 40 ounces (1.1 kg) Porterhouse steaks
- 1 teaspoon Traeger Coffee Rub

Direction:

1. Place the butter, Dijon, and Worcestershire sauce in a small bowl and set aside.

2. Rub the steaks with Traeger Coffee Rub and allow them to sit for 15 minutes.

3. Add butter mixture, spreading evenly on both sides of the steak.

4. Preheat a grill or pan over medium-high heat.

5. Cook the steaks for 45 minutes until the desired doneness.

6. Allow to rest for 5 minutes then slice against the grain and serve.

Nutrition:

- **Energy (calories):** 1456 kcal
- **Protein:** 115.87 g
- **Fat:** 105.87 g
- **Carbohydrates:** 3.62 g
- **Calcium:** 136 mg
- **Magnesium:** 62 mg
- **Phosphorus:** 1088 mg
- **Iron:** 11.82 mg

3. Grilled Bloody Mary Flank Steak

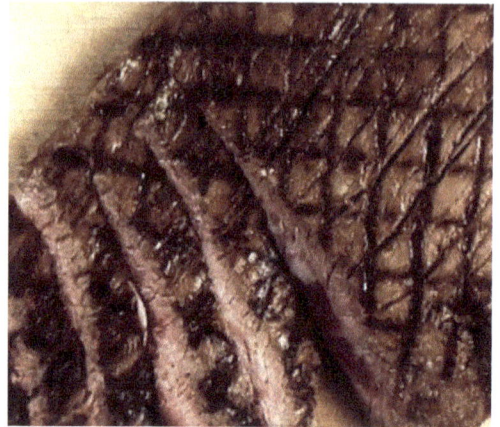

Preparation time: 8 hours

Cooking time: 15 minutes

Servings: 4

Ingredients:

- 1 and one-halves pounds (680 g) flank steak

Marinade:

- 2 cups Traeger Smoked Bloody Mary Mix
- One-half cup vodka
- One-half cup vegetable oil
- 3 clove garlic, minced
- 1 whole lemon or lime, juiced
- 1 tablespoon Worcestershire sauce

- 1 teaspoon celery salt
- 1 teaspoon coarse ground black pepper
- Hot sauce, to taste

Direction:

1. Mix all the ingredients together in a shallow dish
2. Add the steak and coat evenly
3. Cover the dish tightly with plastic wrap
4. Marinate for 8 hours in the refrigerator
5. Remove steak from the marinade and allow excess marinade to drip off
6. Preheat your grill with the lid down
7. When the grill is hot, grill the steak for 15 minutes
8. Cut the steak into 1-inch strips. Enjoy it!

Nutrition:

- **Energy (calories):** 453 kcal
- **Protein:** 25.29 g
- **Fat:** 33.36 g
- **Carbohydrates:** 14.33 g
- **Calcium:** 66 mg
- **Magnesium:** 33 mg
- **Phosphorus:** 283 mg
- **Iron:** 2.07 mg
- **Cholesterol:** 69 mg

4. Smoked Tomahawk Steak

Preparation time: 5 minutes

Cooking time: 1 hour

Servings: 4

Ingredients:

- 1 (32 ounces/907 g) bone-in Tomahawk rib-eye steak, 2 inches thick
- Kosher salt
- Meat Church Holy Cow BBQ Rub
- 3 tablespoons butter

Direction:

1. Season the steak with salt and generously rub with the Holy Cow Rub.

2. Smoke the steak with pellet wood for 1 hour at 250 °F (121 °C).

3. Set a cast-iron skillet over medium-low heat.

4. Melt butter in the skillet and fry steak to the desired doneness. Remove from heat and serve.

Nutrition:

- **Energy (calories):** 373 kcal
- **Protein:** 20.52 g
- **Fat:** 32.53 g
- **Carbohydrates:** 0.01 g
- **Calcium:** 14 mg
- **Magnesium:** 21 mg
- **Phosphorus:** 152 mg
- **Iron:** 1.97 mg

Smoke Pork Recipes

5. Light and Lean Loin

Preparation time: 10 minutes

Cooking time: 5-6 hours

Servings: 4

Ingredients:

- 2 garlic cloves, minced
- 2 teaspoons chili powder
- 2 teaspoons dried oregano
- 2 teaspoons onion powder
- 2 teaspoons ground cumin
- 1 teaspoon salt
- 1 teaspoon freshly ground black pepper

- 2 tablespoons olive oil
- 1 (2 and one-half to 3-pounds) pork loin roast
- One-half cup chicken broth
- One-half cup apple cider vinegar
- 1 cup Alabama White Sauce, for serving

Direction:

1. Preheat the oven to 325 degrees.
2. In a mixing bowl, stir together the dry ingredients. Add the olive oil and mix until combined.
3. Rub the spice mixture over all surfaces of the pork roast, and let the meat sit at room temperature for about 30 minutes.
4. Stir together the broth and the vinegar in a small bowl.
5. Put the pork roast in a 9-by 13-inch baking dish, and pour the broth mixture over the meat. Cover the dish with aluminum foil, and bake for about 5 to 6 hours, until the meat is very tender and ready to be pulled apart.
6. Poke the roast all over with the prongs of a fork, and push the meat back into its original shape. Transfer to a cutting board, cover with foil, and rest for at least 10 minutes.
7. Shred the meat with two forks. Coat the meat with Alabama White Sauce, and serve.

Nutrition:

- **Energy (calories):** 589 kcal
- **Protein:** 59.25 g - **Fat:** 32.67 g
- **Carbohydrates:** 12.32 g
- **Calcium:** 119 mg
- **Magnesium:** 71 mg
- **Phosphorus:** 550 mg
- **Iron:** 2.84 mg
- **Cholesterol:** 175 mg
- **Sugars total:** 6.34 g

6. Beer-smoked Brats

Preparation time: 20 minutes

Cooking time: 1 to 2 hours

Servings: 12

Ingredients:

- 2 and one-half to 3 pounds brats, uncooked (12 links)
- 3 (12-ounces) cans dark beer
- 2 tablespoons unsalted butter, melted
- 2 garlic cloves, minced
- 1 teaspoon freshly ground black pepper
- 1 large onion, sliced
- 2 Poblano peppers, seeded and cut into strips
- 12 sub (hoagie) rolls, for serving

Direction:

1. Beer-Smoked Brats — Using a sharp knife, prick sausages 4 times; set aside. In a small bowl, combine beer, butter, garlic, and pepper; set aside.

2. Preheat a grill to medium heat.

3. Place sausages on a grill, cover, and cook for 1-2 hours, turning occasionally. During the last few minutes of cooking, baste sausages with beer mixture. Remove from grill.

4. Add onions and Poblano peppers to grill and cook until tender, about 10 minutes, turning once.

5. To serve, slice sausages into 1-inch pieces. Place equal portions of sausages and peppers on rolls; serve with additional beer mixture, if desired.

Nutrition:

- **Energy (calories):** 358 kcal
- **Protein:** 7.51 g - **Fat:** 3.28 g
- **Carbohydrates:** 73.62 g
- **Calcium:** 124 mg
- **Magnesium:** 43 mg
- **Phosphorus:** 120mg
- **Iron:** 3.62 mg
- **Cholesterol:** 2 mg
- **Sugars total:** 43.99 g
- **Starch:** 17.08 g

7. Carolina Hash

Preparation time: 15 minutes

Cooking time: 2 hours

Servings: 20

Ingredients:

- 1 tablespoon canola oil
- 3 large onions, chopped
- 2 large potatoes, chopped
- 2 and one-half pounds Beyond Basic Boston Butt, chopped
- 1 (14.5-ounces) can chicken stock
- 2 cups of water
- 2 tablespoons freshly ground black pepper
- 2 teaspoons ground cayenne pepper
- 1 cup Bluesy Competition BBQ Sauce
- One-half cup mustard
- One-half cup apple cider vinegar

- One-fourth cup ketchup
- 2 tablespoons Worcestershire sauce
- One-half cup (1 stick) butter

Direction:

1. Sear the meat in a hot pan until brown. Add 2 tablespoons water to the pan to remove any excess fat. Move to the Traeger grill smoker.

2. Add the onion, potatoes, stock, water, black pepper, cayenne pepper, BBQ sauce, mustard, vinegar, ketchup, and Worcestershire sauce. Cook on high for two hours.

3. Add the butter and serve.

Nutrition:
- **Energy (calories):** 322 kcal
- **Protein:** 18.83 g - **Fat:** 19.59 g
- **Carbohydrates:** 19.77 g
- **Calcium:** 66 mg
- **Magnesium:** 43 mg
- **Phosphorus:** 206 mg
- **Iron:** 2.7 mg
- **Cholesterol:** 81 mg
- **Fiber:** 3.27 g

8. Polynesian Sauce–stuffed Pork Chops

Preparation time: 20 minutes

Cooking time: 1 and one-half hours

Servings: 4

Ingredients:

- 1 cup French dressing
- One-half cup honey
- 1 tablespoon apple cider vinegar
- 4 ounces cream cheese, at room temperature
- One-fourth cup onion, finely chopped
- One-fourth cup celery, finely chopped
- 2 garlic cloves, minced
- 3 teaspoons salt, divided
- 3 teaspoons freshly ground black pepper, divided
- 4 (1-inch-thick) center-cut boneless pork chops

Direction:

1. In a small bowl, stir together the French dressing and honey. Stir in vinegar, cream cheese, onion, celery, garlic, 1 teaspoon salt, and 1 teaspoon pepper. Trim fat from chops. Spoon 3 tablespoons of the dressing mixture onto each of the chops to within ½ inch of edges. Roll up chops, and secure with toothpicks. Sprinkle with remaining 2 teaspoons salt and remaining 1 teaspoon pepper.

2. Place chops in a 13x9-inches glass baking dish. Pour remaining dressing mixture over chops; cover, and refrigerate 1 hour or overnight, turning chops several times.

3. Preheat oven to 325 °F. Bake chops, covered, 1 hour or until cooked through, basting occasionally with pan juices. Transfer chops to platter; drizzle with pan juices. Serve chops with sauce.

Nutrition:

- **Energy (calories):** 655 kcal
- **Protein:** 22.48 g - **Fat:** 41.92 g
- **Carbohydrates:** 50.79 g
- **Calcium:** 65 mg
- **Magnesium:** 32 mg
- **Phosphorus:** 240 mg
- **Iron:** 1.93 mg
- **Cholesterol:** 74 mg
- **Sugars total:** 48.01 g

9. Chinese-Style sticky red Ribs

Preparation time: 5 minutes and 4 hours to overnight

Cooking time: 6 hours

Servings: 8

Ingredients:

- One-half cup hoisin sauce
- One-half cup soy sauce
- 4 garlic cloves, minced
- 3 tablespoons of rice wine vinegar
- 3 tablespoons firmly packed brown sugar
- 3 tablespoons honey
- 1 teaspoon Chinese five-spice powder
- 1 teaspoon red food coloring (optional)
- 8 pork belly spareribs
- One-third cup chopped scallions, both white and green parts, for serving

Direction:

1. Marinate spareribs in the hoisin sauce, soy sauce, garlic, rice wine vinegar, sugar, Chinese five spice powder, and red food coloring (if using) in a large bowl overnight. Transfer ribs to a slow cooking dish. Cover and let sit at room temperature for 30 minutes.

2. Disperse food coloring evenly over the ribs. Cover and refrigerate ate for at least 4 hours and up to overnight.

3. Preheat oven to 300 degrees. Transfer ribs to a dish just large enough to hold them snugly in a single layer. Cover and refrigerate until ready to bake.

4. Bake the ribs, uncovered, until they have browned and almost fallen off the bone, about 6 hours. Cool 10 minutes.

5. Skin the cooled ribs and serve with scallion on the top.

Nutrition:

- **Energy (calories):** 1714 kcal
- **Protein:** 97.72 g
- **Fat:** 140.44 g
- **Carbohydrates:** 15.71 g
- **Calcium:** 103 mg
- **Magnesium:** 96 mg
- **Phosphorus:** 770 mg
- **Iron:** 7.05 mg
- **Cholesterol:** 476 mg
- **Fiber:** 1 g
- **Sugars total:** 11.05 g

Smoke Lamb Recipes

10. Pistachio crusted Lamb with Vegetables

Preparation time: 10 minutes

Cooking time: 3 Hours

Servings: 2

Ingredients:

Pistachio Crusted Lamb:

- 2 racks of lamb
- 1 teaspoon herbs De Provence
- Salt
- Ground black pepper
- 1 tablespoon vegetable oil
- 2/3 cups pistachio nuts, chopped
- 2 tablespoon dry bread crumbs

- 1 tablespoon butter, melted
- 1 teaspoon olive oil
- 3 tablespoon Dijon mustard

Roasted Tri-Color Carrots and Fingerling Potatoes:

- 1 lb. fingerling potato medley
- 1 bunch tri-color carrots, peeled & chopped
- 1 tablespoon olive oil
- 1/2 teaspoon kosher salt
- 1/2 teaspoon ground black pepper
- 1 clove garlic, minced
- 2 teaspoon fresh time, minced

Directions:

1. Preheat the oven to 400 °F.

2. Drizzle the olive oil over the fingerling potatoes and the carrots. Spread them out into a single layer on a large baking sheet and sprinkle salt and pepper over them. Bake the potatoes and carrots for 25-30 minutes, or until they are cooked through.

3. Set a rack to the highest setting in the oven. Add the lamb to the rack and cook it for 3 hours

4. Wait 5 minutes before turning the oven off and opening the door. Remove the lamb from your oven and allow it to sit for 5 minutes before cutting it into slices.

5. Chop the pistachio nuts and mix them together with the bread crumbs, melted butter, and herbs de Provence.

6. Add the seasoned pistachios on top of the lamb rack. Cook the lamb for an additional 30 minutes in the oven.

7. The potatoes will be cooked through during this time.

8. Slice your lamb into portions and serve them topped with your potatoes and carrots.

9. Drizzle the veggies with Dijon mustard before serving.

Nutrition:

Energy (calories): 2614 kcal

Protein: 154.51 g

Fat: 181.81 g

Carbohydrates: 118.26 g

Calcium: 413 mg

Magnesium: 491 mg

Phosphorus: 2319 mg

Iron: 19.7 mg

11. Smoked Lamb Leg with Salsa Verde

Preparation time: 10 minutes

Cooking time: 2 hours

Servings: 2

Ingredients:

Leg of Lamb:

- 1 leg of lamb, aitchbone removed, fat trimmed to 1/4 inch thick, and tied
- 1 head of garlic, peeled
- 2 tablespoon Kosher salt
- 2 tablespoon fresh rosemary, chopped
- 1 teaspoon fresh ground black pepper
- 1/4 dry red wine, or beef broth

Green Garlic Salsa Verde:

- 6 green garlic cloves, unpeeled
- 1 tablespoon capers
- 1 lb. fresh tomatillos, husked, rinsed
- 1 small onion, quartered
- 5 serrano chilies
- 1/4 cup fresh cilantro, chopped
- 1 teaspoon sugar
- kosher salt to taste
- 2 tablespoon olive oil
- 1 cup low salt chicken broth
- 3 tablespoons squeezed lime juice

Directions:

1. Put your lamb in a pan with the garlic head, salt, Rosemary, pepper, and wine.

2. Let it sit. In the mixing bowl, sauté your green garlic in olive oil over medium-high heat until tender. 3. Add the tomatillos, onions, chilies, and cilantro to the bowl with the garlic.

3. After 10 minutes, pour in a quarter cup of water and the sugar. After 20 minutes, add in the lime juice. Add the broth and simmer for 15 minutes. Place the lamb in moderate heat for 2 hours, basting it occasionally with the sauce.

4. Serve warm, garnished with julienned green garlic.

Nutrition:

- **Energy (calories):** 1172 kcal
- **Protein:** 151.16 g
- **Fat:** 53.06 g
- **Carbohydrates:** 15.61 g
- **Calcium:** 118 mg
- **Magnesium:** 211 mg
- **Phosphorus:** 1456 mg
- **Iron:** 14.11 mg

12. Armenian Style Lamb Shanks with Barley Risotto

Preparation time: 10 minutes

Cooking time: 3 hours

Servings: 2

Ingredients:

- 2 tablespoons Pomegranate Molasses
- 1/4 cup tomato paste
- 1 tablespoon garlic powder
- 1 tablespoon ground cinnamon
- 1 teaspoon ground fenugreek
- 1 teaspoon ground cumin
- 1 teaspoon ground cayenne
- 1 teaspoon ground turmeric
- 3 tablespoons + 2 teaspoons kosher salt
- 4 lamb hind shanks
- 2 oz. lamb or beef stock

- 1 cup pearled barley
- 2 tablespoons olive oil
- 1 medium yellow onion, small dice
- 1/2 cup grated Parmigiano Reggiano
- 2 tablespoons butter

Directions:

1. Mix the lamb shanks in a large bowl with 2 tablespoons salt, 1 tablespoon black pepper, and 5 tablespoons olive oil. Cover and refrigerate for at least 4 hours.

2. Preheat oven to 250 degrees F. Toss shanks with the tomato paste and remaining spices to coat.

3. Remove the shanks from their marinade. Heat an oven-proof Dutch oven or large, heavy-bottomed pot over high heat.

4. Add 2 tablespoons of olive oil. Once the oil swirls, cook each shank for 5 minutes. Transfer to a plate and set aside.

5. Add the onions and cook until golden brown. Add the Italian parsley. Cook for 1 minute.

6. Deglaze the pot with 1/4 cup stock and cook for 1 minute. Return the shanks to the pot and add the remaining 4 cups stock, 2 tablespoons kosher salt, and the cooked barley.

7. Don't crowd the pot: if necessary, cook the shanks in two batches.

8. Cover the pan and transfer to the oven. After 4 hours, the meat should start to fall off the bone and the barley should be tender.

9. If the barley is not cooked, uncover, and cook for an additional 15 minutes.

10. Remove the shanks from the pot and serve with a generous amount of grated cheese and a drizzle of molasses.

Nutrition:

- **Energy (calories):** 959 kcal
- **Protein:** 25.69 g
- **Fat:** 33.74 g
- **Carbohydrates:** 150.47 g
- **Calcium:** 351 mg
- **Magnesium:** 154 mg
- **Phosphorus:** 553 mg
- **Iron:** 7.55 mg

13. Juicy Mint Pulled smoked Lamb Shoulder

Preparation time: 6 hours

Cooking time: 6 hours and 10 minutes

Servings: 10

Ingredients:

- Boneless Lamb Shoulder (6-lbs., 2.7-kg.)

The Rub:

- Garlic powder – 2 tablespoons
- Onion powder – 2 tablespoons
- Nutmeg – one-half teaspoon
- Salt – 1 teaspoon
- Pepper – one-half teaspoon

The Liquid:

- Brown sugar – ¼ cup
- Beer – one-fourth cup
- Chopped mint leaves – ¼ cup
- Apple cider vinegar – 1 and one-half tablespoons
- Salt – ¼ teaspoon
- Pepper – one-half teaspoon

Direction:

1. In a large bowl, add the rub ingredients together. Mix well.

2. Rub the meat with the mix on all sides. Add the liquid mixture into the bowl. Mix well. Re-use the same rub mixture. Rub the entire lamb until it is covered. Place in a zip bag and put it in the refrigerator overnight.

3. Heat a slow cooker on low. Place the lamb in the pot and cover.

4. Cook for 6 hours (approximately) until tender. Remove the lamb shoulder from the slow cooker. Shred with forks. Add back to the liquid in the chicken.

5. Add the mustard, molasses, and black pepper. Stir to coat.

6. Reverse sear the pulled lamb. Transfer to a bowl and serve with BBQ sauce on the side.

Nutrition: Energy (calories): 189 kcal, **Protein:** 20.93 g - **Fat:** 6.97 g, **Carbohydrates:** 9.44 g, **Calcium:** 20 mg, **Magnesium:** 28 mg, **Phosphorus:** 194 mg, **Iron:** 45 mg

Smoked Poultry Recipes

14. Smoked Chicken Tikka Drumsticks

Preparation time: 12 hours

Cooking time: 50 minutes

Servings: 6

Ingredients:

- 1 tablespoon smoked paprika
- 1 tablespoon Garam masala
- 1 tablespoon ground cumin
- 1 tablespoon ground coriander
- 1 teaspoon turmeric
- 1 teaspoon ground cayenne pepper
- 1/2 medium yellow onion, diced

- 1 whole ginger, thumb-sized peeled and roughly chopped
- 6 clove garlic, chopped
- 1/2 cup yogurt, Greek
- 1/2 whole lemon, juiced
- 1/4 cup extra-virgin olive oil
- 12 whole chicken drumsticks
- 3/2 cup yogurt, Greek
- 1 tablespoon curry powder
- 1 tablespoon lime juice
- 1 pinch coarse sea salt
- 1 to taste cilantro
- 1/2 small red onion, thinly sliced
- 1 whole lime, cut into wedges
- 2 whole green or red chilies, sliced

Direction:

1. Preheat a smoker or grill to 250 degrees F.

2. In a medium bowl, whisk together the paprika, Garam masala, cumin, coriander, turmeric, cayenne, onion, ginger, garlic, yogurt, lemon juice, olive oil, and a generous pinch each of salt and pepper. Stir in 1 cup of the chicken stock and set the tandoori mixture aside.

3. Place the drumsticks in a large resealable plastic bag, then add the tandoori mixture and toss well. Marinate in the refrigerator for at least 4 hours or, preferably, overnight.

4. Remove the chicken from the marinade and discard the marinade. Pat each drumstick dry with paper towels.

5. Arrange the chicken on the grill and cook for 2 to 3 hours, basting occasionally with the reserved tandoori mixture until the drumsticks develop a nice char and the meat is tender and cooked through. Note: Add wood chips to the fire f

6. or a smokier flavor.

7. To make the yogurt barbecue sauce, place all of the ingredients in a food processor, and blend until smooth.

8. To cook the rice, bring the remaining 3/4 cup chicken stock to a boil in a medium saucepan, then reduce the heat to low and add the rice.

9. Cook, stirring occasionally, for 10 to 15 minutes, or until the rice fluffs and absorbs the stock.

10. Cover the pot with a lid and set aside until the rice is tender, then remove from the heat and fluff the rice with a fork.

11. Serve the drumsticks on the rice and garnish with thinly sliced red onion, lime wedges, and whole chilies. Serve with the yogurt barbecue sauce on the side.

Nutrition: Energy (calories): 620 kcal, **Protein:** 51.86 g, **Fat:** 31.29 g, **Carbohydrates:** 32.52 g, **Calcium:** 161 mg, **Magnesium:** 87 mg,, **Phosphorus:** 541 mg, **Iron:** 3.84 mg

15. Peruvian Roasted Chicken with Green Sauce

Preparation time: 30 minutes

Cooking time: 1 hour

Servings: 6

Ingredients:

- 1 lemon, juiced
- 1 quart cold water
- 4 pound whole chicken
- 2 lime, juiced
- 3 clove garlic, minced
- 1 tablespoon Aji Amarillo paste
- 2 teaspoon huHcatay paste
- 1/4 cup vegetable oil
- 1 tablespoon paprika

- 2 teaspoon granulated sugar
- 2 teaspoon ground cumin
- 3/2 teaspoon kosher salt
- 1 teaspoon freshly ground black pepper
- 4 whole green onion
- 4 romaine lettuce leaves
- 2 clove garlic, chopped
- 2 jalapeño, seeded and chopped
- 1 lime, juiced
- 1/2 cup cilantro leaves
- 3/4 cup mayonnaise
- 1/4 cup sour cream
- 1 tablespoon Aji Amarillo paste
- 4 fresh mint leaves
- To taste salt and pepper

Direction:

1. Wash chicken inside and out with cold water and pierce chicken all over with a sharp knife. Release the breast skin over the breasts and thighs, and place the chicken in a roasting pan. In a small bowl, whisk together the two kinds of garlic, paprika, sugar, cumin, salt and pepper, and 3 Tablespoons of the lime/lemon juice.

2. Rub the mixture all over the chicken, under the skin as well as outside. Tie the legs, season again with salt and pepper, and place it in a 375-degree oven.

3. Roast for 25 minutes, then lower the oven temperature to 325 degrees. Continue to roast until the chicken is dark gold and very aromatic, about 1 hour. Allow the chicken to rest for 10 minutes before carving. To make the Green Sauce: Place the tomatillos, onion, garlic, and green onion into a medium saucepan and cover with water. Bring to boil, reduce heat, and simmer until clear in color and very tender.

4. Drain and rinse under cold water. Place tomatillos, onion, garlic, and green onion into a bowl and coarsely chop. Add the remaining three lime juices and the cilantro, Aji, and salt and pepper.

5. Pour into a blender and blend until smooth. Transfer to the bowl and add mayonnaise, sour cream, and mint. Taste for salt and pepper. Serve immediately as the sauce will discolor.

Nutrition:
- **Energy (calories):** 1308 kcal
- **Protein:** 146.63 g
- **Fat:** 43.41 g
- **Carbohydrates:** 82 g
- **Calcium:** 392 mg
- **Magnesium:** 303 mg
- **Phosphorus:** 1623 mg
- **Iron:** 13.56 mg

16. Roasted Tingle Wings

Preparation time: 10 minutes

Cooking time: 30 minutes

Servings: 6

Ingredients:

- 3 whole jalapeño
- 1 tablespoon Trappey's Red Devil Cayenne Pepper Sauce
- 1/2 cup Traeger Texas Spicy BBQ Sauce
- 2 tablespoon Traeger Blackened Saskatchewan Rub
- 1/2 cup honey
- 1 tablespoon Worcestershire sauce
- 1/4 cup water

Direction:

1. Remove the skins from all ingredients
2. Chop the roasted garlic
3. Blend all ingredients except cayenne pepper sauce in the blender to create the marinade. (you can use a food processor)
4. Place the wings in the marinade.
5. Refrigerate for two to three days.
6. Grill/broil wings until they are brown and crisp.
7. Remove the wings and pour in the cayenne pepper sauce.
8. Broil on high for four to five minutes.
9. Remove from broiler/grill and coat with BBQ sauce.
10. Broil on high for another five minutes.
11. Remove wings from grill/broil and coat with roasted garlic (one tablespoon)
12. Add a dash of blackened Saskatchewan rub (one-half tablespoon).
13. Pour in the honey.
14. Once wings have cooled, slice them up, and enjoy the spicy garlic flavor.
15. Add a dash of blackened Saskatchewan rub (one-half tablespoon).
16. Pour in the honey.
17. Once wings have cooled, slice them up, and enjoy the spicy garlic flavor.

Nutrition:

- **Energy (calories):** 97 kcal
- **Protein:** 0.54 g
- **Fat:** 0.21 g
- **Carbohydrates:** 25.81 g
- **Calcium:** 14 mg
- **Magnesium:** 6 mg
- **Phosphorus:** 13 mg
- **Iron:** 0.42 mg

17. Grilled Asian Lettuce Wrap Turkey Burger

Preparation time: 15 minutes

Cooking time: 15 minutes

Servings: 4

Ingredients:

- 3 tablespoon mayonnaise
- 1 tablespoon Sriracha
- Pound turkey, ground
- 1/2 cup panko breadcrumbs
- 1/4 cup cilantro, finely chopped
- 3 cloves garlic, minced
- 2 scallions, chopped
- 3 tablespoons soy sauce
- 1 tablespoon sesame oil
- 1 tablespoon grated ginger
- 1 teaspoon red pepper flakes

- 1 head Boston lettuce
- 1 tomato, sliced
- 1 red onion, sliced

Direction:

1. Mix together soy sauce, mayonnaise, garlic, sriracha, and sesame oil in a large bowl; season to taste.

2. In another bowl, mix panko crumbs, cilantro, sesame seeds, ginger, and pepper flakes. Shape them into a 1-inch thick patty.

3. Grill patties until the outside is crispy and cooked through, about 15 minutes, flipping halfway through.

4. Top each patty with lettuce leaves, tomato, and onion. Spread 1 tablespoon of the soy sauce mixture on top of each lettuce leaf.

5. Serve as is, or if desired, wrap the whole thing together.

Nutrition:

- **Energy (calories):** 343 kcal
- **Protein:** 17.54 g
- **Fat:** 27.38 g
- **Carbohydrates:** 6.23 g
- **Calcium:** 199 mg
- **Magnesium:** 36 mg
- **Phosphorus.** 173 mg
- **Iron:** 2.68 mg

18. Roasted Tinfoil Dinners

Preparation time: 30 minutes

Cooking time: 30 minutes

Servings: 4

Ingredients:

- 1/2 pound new potatoes, halved
- 1 pint cremini mushrooms, cleaned and halved
- 4 whole chicken breast
- 1 to taste salt and pepper
- 1/2 pound green beans, trimmed
- 1 whole lemon, cut into 3/4" slices
- 1 to taste salt and pepper

Direction:

1. Preheat oven to 450 °F
2. Season chicken breasts with salt and pepper, wrap them in a

sheet of tinfoil.

3. Wrap potatoes in the same sheet of tinfoil, sprinkle salt, pepper, and pour in some water to the sheet.

4. Wrap the mushrooms and the beans in the same tinfoil and then with the warm tinfoil of the potatoes.

5. Clean the chicken breast well and place the slices of lemon in each one of them.

6. When everything is ready, layer the tinfoil packages in the pan. Put the chicken breasts in the middle, then the potatoes to cover.

7. Cover the bottom of the pan with water and bake everything for 30 minutes.

8. Place the food in the center of the table and remove the tinfoil, it's ready to be served.

Nutrition:
- **Energy (calories):** 549 kcal
- **Protein:** 61.98 g - **Fat:** 26.95 g
- **Carbohydrates:** 11.73 g
- **Calcium:** 43 mg
- **Magnesium:** 91 mg
- **Phosphorus:** 545 mg
- **Iron:** 2.81 mg

19. BBQ Chicken baked Potato

Preparation time: 10 minutes

Cooking time: 1 hour

Servings: 4

Ingredients:

- 4 russet potatoes
- 2 chicken breast
- 2 clove garlic, minced
- 2 tablespoon extra-virgin olive oil
- 2 teaspoon Worcestershire sauce
- As needed Traeger pork & poultry rub
- As needed Traeger apricot BBQ sauce
- To taste salt and pepper
- To taste butter
- To taste sour cream
- To taste cheese
- To taste chives, chopped

Direction:

1. Preheat oven to 375 degrees.

2. Scrub potatoes clean and pat dry with paper towels.

3. Rub olive oil onto the tops of potatoes and then sprinkle Traeger Pork & Poultry Rub on top.

4. Put potatoes into the Traeger BBQ and cook for 45 minutes or until done. Poke potatoes with a fork to check doneness, cook longer if necessary.

5. Meanwhile, in a large sauté pan on the grill, toss chicken with garlic, olive oil, and Worcestershire sauce.

6. Once the chicken is browned and the internal temperature reads 165 degrees, transfer to a cutting board and let rest for 10 minutes.

7. Cut potatoes in half.

8. Slice chicken.

9. Assemble all ingredients on a plate, so that people can add the amount of each they like.

10. Serve with Traeger Apricot BBQ Sauce on the side.

Nutrition: Energy (calories): 585 kcal, **Protein:** 38.54 g, **Fat:** 16.75 g, **Carbohydrates:** 70.96 g, **Calcium:** 73 mg, **Magnesium:** 126 mg, **Phosphorus:** 467 mg, **Iron:** 4.59 mg

20. Grill-fried Buttermilk Chicken with Peppercorn Gravy

Preparation time: 10 minutes

Cooking time: 30 minutes

Servings: 8

Ingredients:

- 2 pounds bone-in chicken pieces
- 2½ cups buttermilk
- One-half teaspoon salt
- 1 teaspoon coarse ground black pepper
- 1 teaspoon dried oregano
- 1 teaspoon dried thyme
- 1 teaspoon smoked paprika
- One-half teaspoon cayenne powder
- One-fourth cup mayonnaise

- 3 cloves garlic, crushed and minced
- One-fourth cup fresh parsley, chopped
- 3 cups crushed butter crackers
- Gravy - One-fourth cup butter
- One-fourth cup flour
- 1 tablespoon coarse ground black pepper
- 1 teaspoon salt - 2½ cups milk
- ½ teaspoon onion powder
- ½ teaspoon garlic powder

Direction:

1. Preheat oven to 425 degrees Fahrenheit. In a skillet, heat half of the butter on the grill over medium heat.

2. Plate chicken in a bowl and pour buttermilk over it; add salt, pepper, oregano, thyme, paprika, and cayenne. Mix well.

3. Skewer the chicken on to the skewers. Cover the skewers with foil and bake in a baking sheet for 10 minutes. Take the chicken out, flip over, and bake for another 10 minutes.

4. Take the chicken out.

5. In the same skillet that the chicken was cooked in, set in the remaining butter. Stir in flour. When it is bubbling, slowly whisk in the milk, pepper, salt, onion powder, and garlic powder. Whisk continuously until the gravy thickens and is bubbling.

6. Add mayonnaise, parsley, and garlic. Pour over chicken.

7. Crush the crackers in a bowl and transfer to a separate bowl

that has been linked with a baking sheet. Place a skewer, chicken, and gravy in the bowl. Cover it with butter crackers and bake for another 5 minutes. After it is done, it will be golden-brown.

8. Enjoy your dish!

Nutrition:

- **Energy (calories):** 580 kcal
- **Protein:** 32.34 g - **Fat:** 39.69 g
- **Carbohydrates:** g
- **Calcium:** 230 mg
- **Magnesium:** 69 mg
- **Phosphorus:** 411 mg
- **Iron:** 2.74 mg

Extra Smoke Recipes

21. Wild Boar

Preparation time: 20 minutes

Cooking time: 6 hours

Servings: 4

Ingredients:

- 1 (4 pounds) wild boar roast
- 2 cups BBQ sauce

Marinade:

- 1 tbsp. chopped fresh thyme
- 1/3 cup honey

- One-fourth cup soy sauce
- One-fourth tsp cayenne pepper
- One-half tsp oregano
- One-fourth cup balsamic vinegar
- One-half tsp garlic powder
- 1 cup apple juice

Directions:

1. Roast the wild boar at 350 °F for approximately 20 minutes or until internal temperature reaches 145 degrees Fahrenheit.

2. While the meat is cooking, combine all ingredients for the marinade in a bowl.

3. Marinate the meat for 6 hours in the refrigerator.

4. Drain the meat from the marinade.

5. Pour the marinade over the roast, cover, and cook on low for 5-6 hours

6. Serve with roasted potatoes.

Nutrition: Energy (calories): 388 kcal, **Protein:** 40.35 g, **Fat:** 5.5 g, **Carbohydrates:** 44.5 g, **Calcium:** 94 m, **Magnesium:** 105 mg, **Phosphorus:** 320 mg, **Iron:** 3.27 mg, **Fiber:** 3.3 g

22. Honey Apricot Smoked Lamb Shank

Preparation time: 1 hour

Cooking time: 3-4 Hours

Servings: 6 Servings

Ingredients:

- 3 pounds of whole lamb shank
- 1 cup of olive oil

Glaze Ingredients:

- 1/2 cup honey
- The ½ cup of orange juice concentrate
- 1/2 cup soy sauce
- 1/2 cup apricot jams
- 1 teaspoon ground nutmeg
- 1/2 teaspoon ground cloves

Directions:

1. Take a large mixing bowl and combine all the glaze ingredients in it.

2. Brush the lamb shank generously with the glaze mixture.

3. Marinate the lamb shank a few hours before cooking.

4. Preheat the smoker grill at a high Temperature until the smoke form.

5. Put the lamb on to the electrical smoker grate and cook for 3-4 hours at 220 degrees. Fahrenheit, or until the internal temperate reaches 150 degrees.

6. After every 30 minutes, baste the lamb shank with the glaze.

7. Enjoy it!

Nutrition:

- **Energy (calories):** 833 kcal
- **Protein:** 48.58 g
- **Fat:** 50.28 g
- **Carbohydrates:** 48.31 g
- **Calcium:** 29 mg
- **Magnesium:** 77 mg
- **Phosphorus:** 470 mg
- **Iron:** 4.84 mg

23. Braised Rabbit and Red Wine Stew

Preparation time: 30 minutes

Cooking time: 2 hours

Servings: 4-6 servings

Ingredients:

- 1 skinless rabbit, chopped into pieces (3-lb, 1.4-kgs)
- Olive oil – 1 tablespoon
- Salted butter – 2 tablespoons
- 1 yellow onion, peeled and chopped
- 1 celery stalk, peeled and chopped
- 1 carrot, peeled and chopped
- 2 garlic cloves, peeled and minced
- Flour – 2 tablespoons
- Chicken broth – 4 cups
- Dry red wine – 1 cup

- 1 thyme sprig
- 2 bay leaves
- Salt and black pepper – to taste
- Crusty baguette – to serve

Directions:

1. Warm the olive oil in a Dutch oven over moderately high heat. Add the rabbit pieces to the pot in batches and cook until browned and golden. Set the meat to one side.

2. Melt the butter in the same pot and add the onion, celery, and carrot. Sauté for 10-12 minutes until soft. Add the garlic and sauté for another 60 seconds.

3. Sprinkle over the flour and stir well to combine, cook for 60 more seconds.

4. Next, pour in the chicken broth and red wine. Return the meat to the pot along with the thyme and bay leaves and bring to a simmer.

5. Cover the Traeger oven with a lid and place it on the grill. Cook for approximately 2 hours until the rabbit is cooked through and tender. Season with salt and pepper to taste.

6. Serve with crusty bread.

Nutrition:Energy (calories): 407 kcal, Protein: 42.07 g - **Fat:** 21.69 g, **Carbohydrates:** 6.56 g, **Calcium:** 40 mg, **Magnesium:** 40 mg, **Phosphorus:** 288 mg, **Iron:** 2.48 mg, **Fiber:** 0.8 g

24. Citrus Smoked Goose Breast

Preparation time: 45 minutes

Cooking time: 3 hours

Servings: 8 servings

Ingredients:

- 8 goose breast halves
- Freshly squeezed orange juice – ½ cup
- Olive oil one-third cup
- Dijon mustard one-third cup
- Brown sugar one-third cup
- Soy sauce one-fourth cup
- Runny honey one-fourth cup
- Dried onion, minced 1 tablespoon
- Garlic powder 1 teaspoon

Directions:

1. In a bowl, combine the marinade ingredients and whisk until combined. Coat the goose with the marinade. Cover the bowl and transfer to the fridge for between 3-6 hours.

2. Transfer the goose to the grill, occasionally brushing with the marinade for the first half an hour, before discarding any excess marinade.

3. Continue cooking until the bird's juices run clear and when using a meat thermometer, registers an internal smoke temperature of 165 °F (74 °C), approximately 10-15 minutes.

Nutrition:

- **Energy (calories):** 1173 kcal
- **Protein:** 158.84 g
- **Fat:** 38.28 g
- **Carbohydrates:** 53.18 g
- **Calcium:** 60 mg
- **Magnesium:** 225 mg
- **Phosphorus:** 1723 mg
- **Iron:** 38.69 mg

25. Maple-Glazed Pheasants

Preparation time: 3 hours

Cooking time: 17 hours

Servings: 6 servings

Ingredients:

- 2 whole pheasants (2.5-lb, 1.1-kgs each)
- Brown sugar – ¼ cup
- Kosher salt – ¼ cup
- Water – 4 cups
- Maple syrup – 2 cups

Directions:

1. First, make the brine. Dissolve the sugar and salt in the water.

2. Arrange the pheasant in a large container and pour over the brine mixture. If the birds are not entirely covered, pour over more water.

3. Chill overnight (8-12 hours).

4. Take the birds out of the liquid and pat dry using kitchen paper. Set aside to dry for an hour.

5. Place the pheasants in the smoker.

6. In the meantime, add the maple syrup to a pan over moderately high heat and boil down until thick and syrupy.

7. After the meat has been smoking for an hour, baste the birds with the maple syrup. Continue to base the meat every half an hour.

8. Enjoy warm or allow to cool.

Nutrition:
- **Energy (calories):** 620 kcal
- **Protein:** 55.36 g - **Fat:** 8.6 g
- **Carbohydrates:** 79.38 g
- **Calcium:** 164 mg
- **Magnesium:** 73 mg
- **Phosphorus:** 542 mg
- **Iron:** 2.92 mg

26. Ultimate Duck Breasts

Preparation time: 5 minutes

Cooking time: 20 minutes

Servings: 6

Ingredients:

- 6 skin-on, boneless duck breasts (7.5-oz. 210-gms each)
- Turbinado sugar –¼ cup
- Kosher salt – one-eight cup
- Garlic powder – 3/4 tablespoon
- Light brown sugar 1/3 cup
- Paprika – 1½ tablespoons
- Onion powder 3/4 tablespoon
- Lemon pepper – 1/2 tablespoon
- Black pepper – 1/2 tablespoon
- Dried thyme – 1/2 tablespoon
- Chili powder – 1 teaspoon
- Cumin – ½ tablespoon

Direction:

1. Rinse the duck breasts and gently pat dry with kitchen paper.
2. Score the fat layer in a crisscross pattern using a sharp knife.
3. Combine all rub ingredients in a small bowl.
4. Flip the duck breasts over so that they are sitting fat-side down and coat the non-fat side liberally with the Prepared rub.
5. Arrange one or two duck breasts at a time on the grill, skin side down, and cook for approximately 5 minutes until a brown crust has developed. Once you have rendered as much fat as possible, turn the duck over. Cook for a few more minutes, until medium-rare.
6. Allow the meat to rest for several minutes before slicing and serving.
7. Serve.

Nutrition:

- **Energy (calories):** 142 kcal
- **Protein:** 6.85 g - **Fat:** 2.76 g
- **Carbohydrates:** 23.62 g
- **Calcium:** 42 mg
- **Magnesium:** 11 mg
- **Phosphorus:** 19mg
- **Iron:** 1.77 mg

Smoked Fish and Seafood Recipes

27. Mussels with Pancetta Aioli

Preparation time: 15 minutes

Cooking time: 30 minutes

Servings: 4

Ingredients:

- Three-fourth cup mayonnaise
- 1 tablespoon minced garlic, or more to taste
- 1 4-ounces slice pancetta, chopped
- Salt and pepper
- 4 pounds mussels
- 8 thick slices of Italian bread
- One-fourth cup good-quality olive oil

Ruddy Shelduck
 de: Rostgans
 fr: Tadorne casarca
 es: Tarro Canelo
 ja: アカツクシガモ
 cn: 赤麻鸭
 Tadorna ferruginea

 www.avitopia.net/bird.en/?vid=201501

♀ adult — Photo W.J.Daunicht

Common Shelduck
 de: Brandgans
 fr: Tadorne de Belon
 es: Tarro Blanco
 ja: ツクシガモ
 cn: 翘鼻麻鸭
 Tadorna tadorna

 www.avitopia.net/bird.en/?vid=201505
 www.avitopia.net/bird.en/?wid=201505

♂ adult — Photo W.J.Daunicht

Garganey
 de: Knäkente
 fr: Sarcelle d'été
 es: Cerceta Carretona
 ja: シマアジ
 cn: 白眉鸭
 Spatula querquedula

 www.avitopia.net/bird.en/?vid=203001
 www.avitopia.net/bird.en/?wid=203001

♂ adult — Photo Ferran Pestana

Northern Shoveler
 de: Löffelente
 fr: Canard souchet
 es: Cuchara Común
 ja: ハシビロガモ
 cn: 琵嘴鸭
 Spatula clypeata

 www.avitopia.net/bird.en/?vid=203010
 www.avitopia.net/bird.en/?wid=203010

♂ adult — Photo W.J.Daunicht

Brant Goose
   de: Ringelgans
   fr: Bernache cravant
   es: Barnacla Carinegra
   ja: コクガン
   cn: 黑雁
Branta bernicla

adult — Photo D.Becker

Mute Swan
   de: Höckerschwan
   fr: Cygne tuberculé
   es: Cisne Vulgar
   ja: コブハクチョウ
   cn: 疣鼻天鹅
Cygnus olor

www.avitopia.net/bird.en/?vid=200801
www.avitopia.net/bird.en/?vid=200801

adult — Photo W.J.Daunicht

Tundra Swan
   de: Zwergschwan
   fr: Cygne siffleur
   es: Cisne Chico
   ja: コハクチョウ
   cn: 小天鹅
Cygnus columbianus

www.avitopia.net/bird.en/?vid=200805

adult — Photo W.D.G.Daunicht

Whooper Swan
   de: Singschwan
   fr: Cygne chanteur
   es: Cisne Cantor
   ja: オオハクチョウ
   cn: 大天鹅
Cygnus cygnus

adult — Photo W.D.G.Daunicht

Species of Birds

Ducks and Geese - *Anatidae*

The family of Ducks and Geese occurs in all continents of the world except in Antarctica. The birds grow up to 30 - 180 cm long and live essentially on the water. The front three toes are webbed, the fourth toe is small and shifted upwards. All species swim, some dive well. Most species fly well, only a few are flightless. However, shortly after the breeding season, the birds adopt simple plumage and shed all flight feathers, so that they are unable to fly for some time. The nests are very diverse: there are nests on the ground, in ground caves, in steep walls and in tree hollows. The clutch comprises 4 to 12 eggs, the incubation lasts between 3 and 5 weeks and the young leave the nest soon after hatching.

Greylag Goose
de: Graugans
fr: Oie cendrée
es: Ansar Común
ja: ハイイロガン
cn: 灰雁
Anser anser

www.avitopia.net/bird.en/?vid=200305
www.avitopia.net/bird.en/?aud=200305

adult

Greater White-fronted Goose
de: Blässgans
fr: Oie rieuse
es: Ansar Careto
ja: マガン
cn: 白额雁
Anser albifrons

adult

Taiga Bean-Goose
de: Waldsaatgans
fr: Oie des moissons
es: Ansar Campestre
ja: ヒシクイ
cn: 豆雁
Anser fabalis

adult

Bird Topography

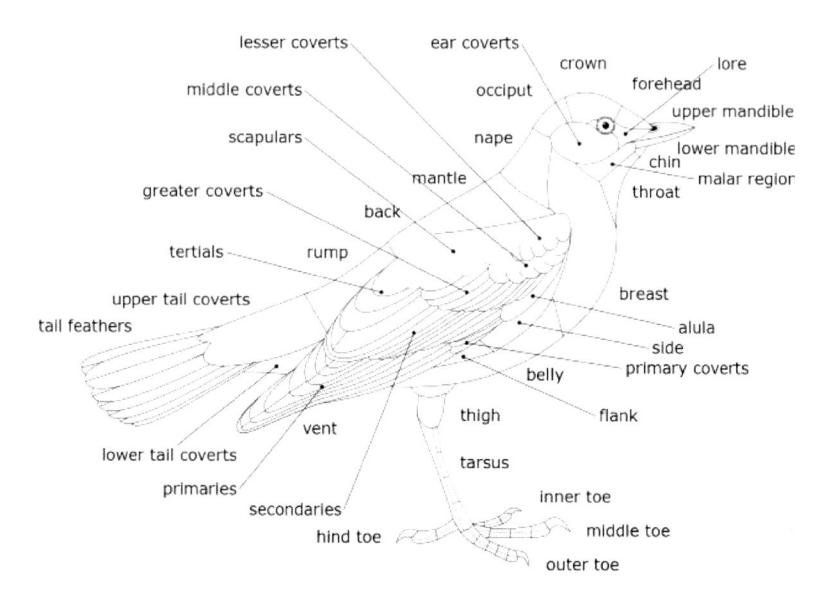

- lesser coverts
- middle coverts
- scapulars
- greater coverts
- tertials
- upper tail coverts
- tail feathers
- lower tail coverts
- primaries
- secondaries
- hind toe
- vent
- rump
- back
- mantle
- ear coverts
- crown
- occiput
- nape
- thigh
- tarsus
- belly
- side
- breast
- throat
- chin
- malar region
- lower mandible
- upper mandible
- forehead
- lore
- alula
- primary coverts
- flank
- inner toe
- middle toe
- outer toe

© AVITOPIA

A map and short descriptions are offered here.

In many cases areas with international designations are declared as National Parks, too, by national administrations.

The assessment of the global conservation status of bird species uses the criteria of the Red List (IUCN) 2012.

Legend

△ Near threatened

▲ Vulnerable

▲ Endangered

▲ Critically endangered

▲ Extinct in the wild

▲ Data deficient

Ø Invalid taxon

† Extinct

ⓔ Picture of an endemic subspecies

🔊 Link to video with audio

▭ Link to video

🔊 Link to audio

This e-book is based on a request to the AVITOPIA Data Base the 18th December 2023.

The request profile was:

   Primary language: English - Secondary language: unrestricted

   Maximum number of pictures per species: 1

   Content and illustration: all names, optimal illustration

   Scientific system: Clements et al. 2017

   Method of area selection: Menu tree

   Name of area: Montenegro

   Survival criterion: unrestricted

   Selection of a taxon: all birds

   Taxonomic depth: Species of Birds

   Selection of activity/nest/portrait: unrestricted

   Selection of plumage/egg(s): unrestricted

   Selection of image technique: unrestricted

In the resulting PDF or ePub file, resp., all index and register entries are linked.

Preface

Montenegro is a souvereign State.

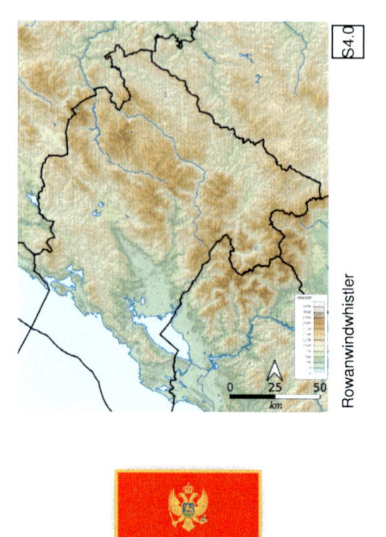

Montenegro lies east of the Adriatic Sea; it is a mountainous, wooded country in the Dinaric Mountains. The altitudes range from a narrow coastal strip on the Mediterranean to Zla Kolata at 2534 m above sea level in the Prokletije massif. In between lies a waterless karst plateau. Only in the southeast are lowlands like those on Lake Skadar. The variety of landscapes makes Montenegro a "hot spot" for biodiversity - the bird life is also extraordinarily diverse for such a small country. Montenegro has protected almost 10% of its area and established five national parks for it. BirdLife International has recognized five areas as IBAs (Important Bird and Biodiversity Areas). Ornithological trips to Montenegro can be very rewarding.

Montenegro contributes to the natural heritage of humanity. You can obtain informations about the importance of this property on the following web page of the UNESCO:

Durmitor National Park - https://whc.unesco.org/en/list/100

Montenegro has joined the UNESCO program to reconcile the conservation of biodiversity with its sustainable use. Biosphere reserves are being set up to promote this purpose. You can find information about these reserves on the following UNESCO website:

Biosphere Reserves - https://en.unesco.org/biosphere/eu-na#montenegro

If you want to assess the importance of individual areas of a country for bird life, the organization BirdLife International is particularly helpful. For this purpose, the terms IBA and EBA were coined.

An IBA (Important Bird and Biodiversity Area) is an area of significant importance for the long-term conservation of global bird life. A list of the IBAs (Important Bird and Biodiversity Areas) of this country is available in the Data Zone of

BirdLife International (IBAs) - https://datazone.birdlife.org/site/results?cty=272

In this list you can call up a map and further information for each IBA with one click.

Many wetlands around the world are protected under the Ramsar Convention to preserve habitat, particularly for waterfowl and shorebirds. A documentation of the country's Ramsar sites can be found on the website

Ramsar wetlands - https://rsis.ramsar.org/sites/default/files/rsiswp_search/exports/Ramsar-Sites-annotated-summary-Montenegro.pdf

Swallows	79
Tits	80
Penduline Tits	82
Long-tailed Tits	83
Nuthatches	83
Wallcreepers	84
Holarctic Treecreepers	85
Wrens	86
Dippers	86
Goldcrests/Kinglets	87
Bush-Warblers and allies	88
Leaf Warblers	88
Reed-Warblers and allies	89
Grassbirds and allies	92
Cisticolas and allies	93
Warblers	93
Flycatchers	95
Thrushes	100
Starlings	102
Accentors	103
Wagtails and Pipits	103
Waxwings	106
Longspurs and Snow Buntings	106
Old World Buntings	107
Finches	109
Sparrows	113
Indices of Names	115
Index of English Names	115
Index of German Names	119
Index of French Names	122
Index of Spanish Names	125
Index of Japanese Names	128
Index of Chinese Names	131
Index of Scientific Names	134
Additional Copyright Terms	138

Conclusion

Congratulations on making it to the end of this book. You can cook anything you can imagine with the Traeger grill smoker., You will find that this Traeger grill smoker is a flexible tool that has good service.

As we all have been able to testify that the use of the pellet grill has been made simple by Traeger: its intuitive control panel has a power button and a knob that allows you to comfortably adjust the temperature. In addition, foods cooked with this grill acquire an incredibly delicious taste. I hope the recipes I gave you may have been simple for you to recreate but delicious at the same time. I am sure that after cooking with my recipes you will have become a real chef. Enjoy.

3. Add the veggies to a baking pan.
4. Roast for 20 minutes.
5. Serve roasted veggies with hummus.

Nutrition:
- **Energy (calories):** 216 kcal
- **Protein:** 6.41 g
- **Fat:** 7.69 g
- **Carbohydrates:** 33.17 g
- **Calcium:** 83 mg
- **Magnesium:** 63 mg
- **Phosphorus:** 157 mg
- **Iron:** 2.07 mg
- **Dietary fiber:** 11.4 g

40. Roasted Veggies & Hummus

Preparation time: 30 minutes

Cooking time: 20 minutes

Servings: 4

Ingredients:

- 1 white onion, sliced into wedges
- 2 cups butternut squash
- 2 cups cauliflower, sliced into florets
- 1 cup mushroom buttons
- Olive oil
- Salt and pepper to taste
- Hummus

Direction:

1. Set the Traeger wood pellet grill.
2. Preheat it for 10 minutes.

Nutrition:

- **Energy (calories):** 41 kcal
- **Protein:** 0.94 g - **Fat:** 2.87 g
- **Carbohydrates:** 3.73 g
- **Calcium:** 26 mg
- **Magnesium:** 23 mg
- **Phosphorus:** 29 mg
- **Iron:** 0.42 mg

39. Grilled Sweet Potato Planks

Preparation time: 30 minutes

Cooking time: 30 minutes

Servings: 8

Ingredients:

- 5 sweet potatoes, sliced into planks
- 1 tablespoon olive oil
- 1 teaspoon onion powder
- Salt and pepper to taste

Direction:

1. Put fire into the Traeger wood pellet grill to high.
2. Preheat it for 15 minutes.
3. Coat the sweet potatoes with oil.
4. Sprinkle with onion powder, salt, and pepper.
5. Grill the sweet potatoes for 15 minutes.

- 3 tablespoons Traeger Leinenkugel's Summer Shandy Rub
- 1 Bunch Dill Weed, fresh
- 12 small Persian cucumbers

Direction:

1. Combine salt, peppercorns, celery seed, coriander, mustard seed, and garlic in a spice bag. Add water and let it rest overnight.

2. Rinse the cucumbers well, discard excess water. Remove the top and bottom end, then cut them in half lengthwise, leaving the peel on. Put the cucumbers in a large pot, add vinegar, beer rub, sugar, and the spice bag.

3. Cook in a slow cooker at 175 degrees on low for 1 hour.

4. Remove spice bag, and using a hand blender, blend the ingredients to the desired consistency.

5. Add fresh dill and season with more vinegar, beer rub, and/or sugar.

6. Serve immediately or refrigerate.

Nutrition:
- **Energy (calories):** 63 kcal
- **Protein:** 2.82 g - **Fat:** 0.78 g
- **Carbohydrates:** 12.57 g
- **Calcium:** 64 mg
- **Magnesium:** 47 mg
- **Phosphorus:** 90 mg
- **Iron:** 1.18 mg

38. Smoked Whole Pickles

Preparation time: 15 minutes

Cooking time: 1 hour

Servings: 8

Ingredients:

- 1/2 cup Kosher salt
- 1/2 teaspoon peppercorns, pink
- 1 teaspoon black peppercorn
- 3/2 teaspoon celery seed
- 3/2 teaspoon coriander seeds
- 1 teaspoon mustard seeds
- 8 cloves garlic, peeled
- 1 quart water
- 1/2 quart white vinegar
- 1/4 cup sugar

Direction:

1. Preheat oven to 350 degrees F. Wash and scrub potatoes. Poke holes with a fork and lay in a flat baking pan. Bake potatoes for 60 minutes or till cooked through and tender. Remove from the oven.

2. In a large pot cook bacon over medium heat till crispy.

3. Remove from pan and drain on paper towels.

4. In the same pot sauté onions in butter until soft, about 3-5 minutes.

5. Add bacon back into the pot and stir to combine.

6. Add chicken stock and milk then bring to a boil stirring often.

7. In a separate pot boil potato over medium heat till soft for about another 15 minutes. Strain potatoes, allowing to cool slightly.

8. Peel the potatoes and mash.

9. Add potatoes and all other ingredients (not sour cream) into the stock and slowly combine.

10. Cook about 20 minutes on low heat. Our garnish is sour cream, bacon, diced onion, and chives lightly sautéed in butter. Ladle into bowls and top with garnish to serve.

Nutrition:

- **Energy (calories):** 813 kcal
- **Protein:** 50.98 g - **Fat:** 31.17 g
- **Carbohydrates:** 84.97 g
- **Calcium:** 317 mg
- **Magnesium:** 151 mg
- **Phosphorus.** 719 mg
- **Iron:** 5.98 mg

37. Smoked Baked Potato Soup

Preparation time: 15 minutes

Cooking time: 1 hour

Servings: 6

Ingredients:

- 6 large russet potatoes
- 12 ounces bacon
- 4 tablespoons butter
- 1 small onion, diced
- 1/4 cup flour
- 4 Cup milk
- 1 can chicken stock
- 1 teaspoon onion powder
- 1 teaspoon garlic powder
- 2 teaspoons salt
- 1 cup sour cream

Nutrition:

- **Energy (calories):** 419 kcal
- **Protein:** 14.05 g
- **Fat:** 36.25 g
- **Carbohydrates:** 12.58 g
- **Calcium:** 454 mg
- **Magnesium:** 94 mg
- **Phosphorus:** 270 mg
- **Iron:** 2.28 mg

- 1/2 cup Parmesan cheese
- 1/2 cup Romano Cheese
- 1/2 cup panko breadcrumbs

Direction:

1. In a large skillet over medium heat, cook and stir the leek until leek have softened and turned tender (about 3 minutes). Add the spinach and stir to combine.

2. Bake, uncovered, at 350° for about 20 minutes or until cheese is melted and spinach is evenly distributed.

3. Roughly chop the spinach. In a large skillet, over medium heat, cook shallot and garlic in butter until tender. Stir in pepper flakes and cook for 1 minute. Add cream, nutmeg, salt, and pepper and bring to a boil, reduce to a simmer and cook for 10 minutes, stirring occasionally. Add sour cream, Romano cheese, Parmesan of and panko breadcrumbs, and stir until cheese is melted and breadcrumbs are golden and crisp. When finished, place the entire mixture in a food processor and pulse until smooth.

4. Add mixture to the macaroni and toss to coat. Cover and refrigerate for about 1 hour before serving.

5. Garnish with grated parmesan cheese and freshly ground pepper.

6. Note: This recipe can be made with fresh chopped spinach.

36. Baked Creamed Spinach

Preparation time: 15 minutes

Cooking time: 30 minutes

Servings: 4

Ingredients:

- 2 tablespoons butter
- 1 finely chopped shallot
- 2 cloves garlic
- 1 teaspoon red pepper flakes
- 3/2 cup heavy cream
- 1 teaspoon ground nutmeg
- To taste salt
- To taste black pepper
- 2 (10 oz.) package frozen spinach, thawed and drained
- 3/4 cup sour cream

Nutrition:

- **Energy (calories):** 240 kcal
- **Protein:** 15.05 g
- **Fat:** 18.32 g
- **Carbohydrates:** 3.01 g
- **Calcium:** 167 mg
- **Magnesium:** 30 mg
- **Phosphorus:** 269 mg
- **Iron:** 3.78 mg

- 1 teaspoon kosher salt
- 1/2 teaspoon black pepper

Direction:

1. preheat oven to 350

2. Grease eight 4-ounces ramekins (or five 6-ounces ramekins) with cooking spray

3. Place oil into a skillet over medium-high heat; add onion and sauté until softened about 2-3 minutes. Take the skillet off heat, transfer onions into a small plate, set aside to cool

4. Purée the spinach in a food processor until smooth, about 2 minutes.

5. Whisk eggs, cheese, puréed spinach, basil, salt, and pepper to a bowl.

6. Pour this mixture into 8 (5-ounces) ramekins.

7. Add onions to the ramekins evenly.

8. Bake for about 15 minutes until the egg is completely cooked and a toothpick comes out clean. Then ready to serve.

Smoked Vegetable Recipes

35. Baked Breakfast Mini Quiches

Preparation time: 15 minutes

Cooking time: 15 minutes

Servings: 8

Ingredients:

- As needed cooking spray
- 1 tablespoon extra-virgin olive oil
- 1/2 yellow onion, diced
- 3 cups Spinach, fresh
- 10 eggs
- 4 ounces shredded Cheddar, mozzarella, or Swiss cheese
- 1/4 cup fresh basil

2. After washing and drying the shrimp, mix it well with the oil and seasonings. Add skewers to the shrimp and set the bowl of shrimp aside.

3. Open the skewers and flip them.

4. Cook for 4 more minutes. Remove when the shrimp is opaque and pink.

Nutrition:
- **Energy (calories):** 93 kcal
- **Protein:** 11.87 g - **Fat:** 4.2 g
- **Carbohydrates:** 1.42 g
- **Calcium:** 87 mg
- **Magnesium:** 22 mg
- **Phosphorus:** 117 mg
- **Iron:** 1.37 mg
- **Cholesterol:** 179 mg

34. Grilled Shrimp

Preparation time: 0 minutes

Cooking time: 15 minutes

Servings: 4

Ingredients:

- Jumbo shrimp peeled and cleaned - 1 lb.
- Oil - 2 tbsp. - Salt – one-half tbsp.
- Skewers - 4-5
- Pepper – one-eight tbsp.
- Garlic salt – one-half tbsp.

Directions:

1. Preheat the wood pellet grill to 375 degrees. Mix all the ingredients in a small bowl.

Nutrition:

- **Energy (calories):** 275 kcal
- **Protein:** 27.37 g
- **Fat:** 14.52 g
- **Carbohydrates:** 7.69 g
- **Calcium:** 172 mg
- **Magnesium:** 58 mg
- **Phosphorus:** 271 mg
- **Iron:** 2.84 mg
- **Fiber:** 1 g

- 1 cup Italian bread crumbs
- One-half cup mayonnaise
- 1 large egg, beaten
- 2 teaspoons fresh parsley, chopped
- 1 and one-half teaspoons salt and pepper

Directions:

1. Take a food processor and add shrimp, chop them up
2. Take a skillet and place it over medium-high heat, add butter and allow it to melt
3. Sauté the onions for 3 minutes
4. Add chopped shrimp with cooled Sautéed onion alongside remaining ingredients listed under stuffing ingredients and transfer to a bowl
5. Cover the mixture and allow it to refrigerate for 60 minutes
6. Rub both sides of the fillet with olive oil
7. Spoon 1/3 cup of the stuffing to the fillet
8. Flatten out the stuffing onto the bottom half of the fillet and fold the Tilapia in half
9. Secure with 2 toothpicks
10. Dust each fillet with smoked paprika and Old Bay seasoning
11. Preheat your smoker to 400 degrees Fahrenheit
12. Transfer to your smoker and smoker for 30-45 minutes.
13. Allow the fish to rest for 5 minutes and enjoy it!

33. Stuffed Shrimp Tilapia

Preparation time: 20 minutes

Cooking time: 45 minutes

Servings: 5

Ingredients:

- 5 ounces fresh, farmed tilapia fillets
- 2 tablespoons extra virgin olive oil
- 1 and one-half teaspoons smoked paprika
- 1 and one-half teaspoons Old Bay seasoning

Shrimp stuffing:

- 1-pound shrimp, cooked and deveined
- 1 tablespoon salted butter
- 1 cup red onion, diced

Gadwall
de: Schnatterente
fr: Canard chipeau
es: Anade Friso
ja: オカヨシガモ
cn: 赤膀鸭
Mareca strepera

www.avitopia.net/bird.en/?vid=203101

♂ adult
Photo D.Dewhurst

Eurasian Wigeon
de: Pfeifente
fr: Canard siffleur
es: Silbón Europeo
ja: ヒドリガモ
cn: 赤颈鸭
Mareca penelope

www.avitopia.net/bird.en/?vid=203103

♂ adult
Photo W.J.Daunicht

Mallard
de: Stockente
fr: Canard colvert
es: Ánade Real
ja: マガモ
cn: 绿头鸭
Anas platyrhynchos

www.avitopia.net/bird.en/?vid=203210
www.avitopia.net/bird.en/?aud=203210

♂ adult
Photo W.J.Daunicht

Northern Pintail
de: Spießente
fr: Canard pilet
es: Ánade Rabudo
ja: オナガガモ
cn: 针尾鸭
Anas acuta

www.avitopia.net/bird.en/?vid=203216

♂ adult
Photo D.Menke

Green-winged Teal
de: Krickente
fr: Sarcelle d'hiver
es: Cerceta común
ja: コガモ
cn: 绿翅鸭
Anas crecca

www.avitopia.net/bird.en/?vid=203219

♂ adult

Marbled Duck
de: Marmelente
fr: Marmaronette marbrée
es: Cerceta Pardilla
ja: ウスユキガモ
cn: 云石斑鸭
Marmaronetta angustirostris
Vulnerable.

www.avitopia.net/bird.en/?vid=203401

adult

Red-crested Pochard
de: Kolbenente
fr: Nette rousse
es: Pato Colorado
ja: アカハシハジロ
cn: 赤嘴潜鸭
Netta rufina

www.avitopia.net/bird.en/?vid=203701
www.avitopia.net/bird.en/?wid=203701

♂ breeding

Common Pochard
de: Tafelente
fr: Fuligule milouin
es: Porrón Europeo
ja: ホシハジロ
cn: 红头潜鸭
Aythya ferina
Vulnerable.

www.avitopia.net/bird.en/?vid=203803

♂ adult

Ferruginous Duck
  de: Moorente
  fr: Fuligule nyroca
  es: Porrón Pardo
  ja: メジロガモ
  cn: 白眼潜鸭
Aythya nyroca
Near threatened.

www.avitopia.net/bird.en/?vid=203805

♂ adult

Tufted Duck
  de: Reiherente
  fr: Fuligule morillon
  es: Porrón Moñudo
  ja: キンクロハジロ
  cn: 凤头潜鸭
Aythya fuligula

www.avitopia.net/bird.en/?vid=203809

♂ adult

Greater Scaup
  de: Bergente
  fr: Fuligule milouinan
  es: Porrón Bastardo
  ja: スズガモ
  cn: 斑背潜鸭
Aythya marila

♂ adult

White-winged Scoter
  de: Samtente
  fr: Macreuse brune
  es: Negrón Especulado
  ja: ビロアドキンクロ
  cn: 斑脸海番鸭
Melanitta fusca
Vulnerable.

♂ adult

Black Scoter
 de: Trauerente
 fr: Macreuse noire
 es: Negrón Común
 ja: クロガモ
 cn: 黑海番鸭
 Melanitta nigra

♂ adult

Common Goldeneye
 de: Schellente
 fr: Garrot à oeil d'or
 es: Porrón Osculado
 ja: ホオジロガモ
 cn: 鹊鸭
 Bucephala clangula

www.avitopia.net/bird.en/?vid=204502

♂ adult

Smew
 de: Zwergsäger
 fr: Harle piette
 es: Serreta Chica
 ja: ミコアイサ
 cn: 斑头秋沙鸭
 Mergellus albellus

www.avitopia.net/bird.en/?vid=204601

♂ breeding

Common Merganser
 de: Gänsesäger
 fr: Grand Harle
 es: Serreta Grande
 ja: カワアイサ
 cn: 普通秋沙鸭
 Mergus merganser

♂ adult

Red-breasted Merganser
  de: Mittelsäger
  fr: Harle huppé
  es: Serreta Mediana
  ja: ウミアイサ
  cn: 红胸秋沙鸭
Mergus serrator

♂ adult

White-headed Duck
  de: Weißkopf-Ruderente
  fr: Érismature à tête blanche
  es: Malvasía Cabeciblanca
  ja: カオジロオタテガモ
  cn: 白头硬尾鸭
Oxyura leucocephala
Endangered.

♂ adult

Pheasants - *Phasianidae*

The pheasant family is distributed worldwide, with the exception of northern Asia, southern South America and the polar regions. The lengths are very different and range from 13 cm to 2 m. They all have round wings, short necks and short thick beaks. The plumage is often very conspicuously patterned and the sexes are mostly different. Most pheasants live on the ground, but some species sleep in trees.

Common Quail
  de: Wachtel
  fr: Caille des blés
  es: Codorniz Común
  ja: ヨアロッパウズラ
  cn: 西鹌鹑
Coturnix coturnix

www.avitopia.net/bird.en/?vid=326802

♀ adult

Rock Partridge
de: Steinhuhn
fr: Perdrix bartavelle
es: Perdiz Griega
ja: ハイイロイワシャコ
cn: 欧石鸡
Alectoris graeca
Near threatened.

adult

Common Pheasant
de: Fasan
fr: Faisan de Colchide
es: Faisán Vulgar
ja: キジ
cn: 环颈雉
Phasianus colchicus

www.avitopia.net/bird.en/?kom=328701
www.avitopia.net/bird.en/?vid=328701
www.avitopia.net/bird.en/?aud=328701

♂ adult

Grey Partridge
de: Rebhuhn
fr: Perdrix grise
es: Perdiz Pardilla
ja: ヨアロッパヤマウズラ
cn: 灰山鹑
Perdix perdix

www.avitopia.net/bird.en/?vid=329101

♂ breeding

Western Capercaillie
de: Auerhuhn
fr: Grand Tétras
es: Urogallo Común
ja: ヨアロッパオオライチョウ
cn: 松鸡
Tetrao urogallus

www.avitopia.net/bird.en/?vid=329302

♂ adult

Black Grouse
 de: Birkhuhn
 fr: Tétras lyre
 es: Gallo-lira Común
 ja: クロライチョウ
 cn: 黑琴鸡
Tetrao tetrix

♂ adult

Hazel Grouse
 de: Haselhuhn
 fr: Gélinotte des bois
 es: Grévol Común
 ja: エゾライチョウ
 cn: 花尾榛鸡
Tetrastes bonasia

♂ adult

Rock Ptarmigan
 de: Alpenschneehuhn
 fr: Lagopède alpin
 es: Lagópodo Alpino
 ja: ライチョウ
 cn: 岩雷鸟
Lagopus muta

♂ breeding

Divers - *Gaviidae*

The species in the family of Loons occur in the higher latitudes of the northern hemisphere, however, they do migrate. They are between 65 cm and 95 cm long. They have small wings, webbed feet and powerful, pointed beaks. They can dive for up to 40 s, but they need a longer approach to take off. Their nests are always created on the shores of freshwater lakes. The young birds are led by both parents.

Red-throated Loon
de: Sterntaucher
fr: Plongeon catmarin
es: Colimbo Chico
ja: アビ
cn: 红喉潜鸟
Gavia stellata

breeding

Black-throated Loon
de: Prachttaucher
fr: Plongeon arctique
es: Colimbo Ártico
ja: オオハム
cn: 黑喉潜鸟
Gavia arctica

breeding

Grebes - *Podicipedidae*

The family of Grebes are found on freshwater lakes around the world, except in the extreme north and south and on some islands. In winter they can also be found on the coast of the sea. The size ranges from 20 cm to 50 cm, the wings are short, tail feathers are missing. The toes have flap-like widenings. They only fly regularly and at night during the migration time. In addition, they are well adapted to aquatic life. Both parents lead the striped or spotted young birds until they become independent.

Little Grebe
de: Zwergtaucher
fr: Grèbe castagneux
es: Zampullín Común
ja: カイツブリ
cn: 小鸊鷉
Tachybaptus ruficollis

www.avitopia.net/bird.en/?kom=375202
www.avitopia.net/bird.en/?vid=375202

breeding

Photo W.J.Daunicht

Horned Grebe
de: Ohrentaucher
fr: Grèbe esclavon
es: Zampullín Cuellirrojo
ja: ミミカイツブリ
cn: 角鸊鷉
Podiceps auritus
Vulnerable.

breeding

Photo D.Dewhurst

Red-necked Grebe
de: Rothalstaucher
fr: Grèbe jougris
es: Somormujo Cuellirrojo
ja: アカエリカイツブリ
cn: 赤颈鸊鷉
Podiceps grisegena

www.avitopia.net/bird.en/?kom=375503
www.avitopia.net/bird.en/?vid=375503

breeding

Photo W.J.Daunicht

Great Crested Grebe
de: Haubentaucher
fr: Grèbe huppé
es: Somormujo Lavanco
ja: カンムリカイツブリ
cn: 凤头䴘䴘
Podiceps cristatus

www.avitopia.net/bird.en/?vid=375504
www.avitopia.net/bird.en/?wid=375504

breeding

Black-necked Grebe
de: Schwarzhalstaucher
fr: Grèbe à cou noir
es: Zampullín Cuellinegro
ja: ハジロカイツブリ
cn: 黑颈䴘䴘
Podiceps nigricollis

breeding

Flamingoes - *Phoenicopteridae*

The family of Flamingos includes only a few species, but they have spread to the warm areas of all continents except Australia. With their long legs and long necks, the birds reach heights of up to 120 cm. The wings are large, the tail is short. The beaks are unique: bent down in the middle with a channel-shaped lower beak and a lid-shaped upper beak. In the water the beak is held upright; the tongue acts as a piston that sucks in the water and presses it out through the filters of the beak. Flamingos feed on small animals and parts of plants that float in the water. They breed in colonies, the nests are truncated cones made of mud. The young birds flee the nest and are fed by both parents.

Greater Flamingo
de: Rosaflamingo
fr: Flamant rose
es: Flamenco Común
ja: ベニイロフラミンゴ
cn: 大红鹳
Phoenicopterus roseus

www.avitopia.net/bird.en/?vid=400103

adult

Petrels - *Procellariidae*

The family of Petrels is at home at sea all over the world. They essentially use the land for breeding, and some species even do so on the coast of Antarctica. Most of the species are migratory birds. The body length ranges from 30 cm to 90 cm. The birds have long, pointed wings and short tails, and their feet are webbed. The smaller species breed in caves or crevices, they defend the young birds by vomiting stinking oil.

Cory's Shearwater
  de: Gelbschnabel-Sturmtaucher
  fr: Puffin cendré
  es: Pardela cenicienta
  ja: オニミズナギドリ
  cn: 猛鹱
Calonectris diomedea

adult — Photo Silveira

Manx Shearwater
  de: Schwarzschnabel-Sturmtaucher
  fr: Puffin des Anglais
  es: Pardela Pichoneta
  ja: マンクスミズナギドリ
  cn: 大西洋鹱
Puffinus puffinus

adult — Photo Martin Reith

Yelkouan Shearwater
  de: Mittelmeer-Sturmtaucher
  fr: Puffin yelkouan
  es: Pardela Mediterránea
  ja: ヒメミズナギドリ
  cn: 地中海鹱
Puffinus yelkouan
Vulnerable.

adult — Photo Emoeke Denes

Storm-petrels - *Hydrobatidae*

The family of Storm-petrels is widespread on all oceans of the earth and occurs partly in large numbers. With a body length of 14 cm to 25 cm, they are the smallest seabirds with webbed feet. Your weak legs are hardly able to carry their body weight without the support of the wings. They breed in colonies in caves or crevices, which they usually only attend at night. Although they usually breed on mammal-free islands, the greatest danger comes from introduced mammals. The Guadalupe storm-petrel was driven to extinction by feral cats.

Wilson's Storm Petrel
de:Buntfuß-Sturmschwalbe
fr: Océanite de Wilson
es:Paíño de Wilson
ja: アシナガウミツバメ
cn:黄蹼洋海燕
Oceanites oceanicus

adult

European Storm Petrel
de:Sturmschwalbe
fr: Océanite tempête
es:Paíño Europeo
ja: ヒメウミツバメ
cn:暴风海燕
Hydrobates pelagicus

adult

Leach's Storm Petrel
de:Wellenläufer
fr: Océanite cul-blanc
es:Paíño Boreal
ja: コシジロウミツバメ
cn:白腰叉尾海燕
Oceanodroma leucorhoa

adult

Storks - *Ciconiidae*

The family of storks is widespread worldwide except for the coldest areas. Some species are resident birds, others are long-distance migrants. The body length ranges from 75 cm to 150 cm. They are long-legged birds with large wings, a long neck and a long beak. They fly a lot, usually with a stretched neck, and they sail excellently. They feed on small animals, from insects to small mammals. 3 to 6 eggs are laid in the shallow nest made of brushwood, which are incubated by both parents.

Black Stork
de: Schwarzstorch
fr: Cigogne noire
es: Cigüeña Negra
ja: ナベコウ
cn: 黑鹳
Ciconia nigra

www.avitopia.net/bird.en/?vid=550201

adult

Photo W.J.Daunicht

White Stork
de: Weißstorch
fr: Cigogne blanche
es: Cigüeña Blanca
ja: シュバシコウ
cn: 白鹳
Ciconia ciconia

www.avitopia.net/bird.en/?kom=550206
www.avitopia.net/bird.en/?vid=550206

adult

Photo W.J.Daunicht

Gannets, Boobies - *Sulidae*

The family of Boobies is common on all seas near the coast. The birds are 65 cm to 100 cm long. The wings are long and pointed, the legs are short and the feet are large and webbed. The beak is strong and has no nostrils. Boobies are extremely elegant fliers, but quite awkward on the ground. They hunt fish for which they plunge into water from a height of up to 35 m in order to pursue them under water and to grab them with their beak. They breed in colonies on the ground or on trees. The young birds are provided with regurgitated food.

Northern Gannet
de: Basstölpel
fr: Fou de Bassan
es: Alcatraz Atlántico
ja: シロカツオドリ
cn: 北鲣鸟
Morus bassanus

♂♀ adult

Cormorants - *Phalacrocoracidae*

The cormorant family are gregarious freshwater or marine birds found on every continent. The body length is 50 cm to 100 cm. They have short legs with large webbed feet. They have a long neck and a slender beak with a curved tip. They dive from the surface of the water and can stay underwater for minutes. The caught fish are thrown into the air and devoured head first.

Pygmy Cormorant
de: Zwergscharbe
fr: Cormoran pymée
es: Cormorán pigmeo
ja: コビトウ
cn: 侏鸬鹚
Microcarbo pygmeus

adult

Great Cormorant
de: Kormoran
fr: Grand Cormoran
es: Cormorán Grande
ja: カワウ
cn: 普通鸬鹚
Phalacrocorax carbo

www.avitopia.net/bird.en/?vid=625206

breeding

European Shag
de: Krähenscharbe
fr: Cormoran huppé
es: Cormorán Moñudo
ja: ヨアロッパヒメウ
cn: 欧鸬鹚
Phalacrocorax aristotelis

♂♀ adult

Pelicans - *Pelecanidae*

The family of Pelicans is scattered across all continents. The body length is 125 cm to 180 cm. The legs are short and strong, the toes are webbed. The wings are large, the tail is short, and the birds sail excellently. The beak is long, straight and flat, and at the lower part of the beak is a large, stretchy pouch that is used like a fishing net. Pelicans are very sociable and often work together to fish. Some species plunge into the water from great heights. Pelicans nest in colonies.

Great White Pelican
de: Rosapelikan
fr: Pélican blanc
es: Pelícano Común
ja: モモイロペリカン
cn: 白鹈鹕
Pelecanus onocrotalus

www.avitopia.net/bird.en/?vid=675104

adult

Dalmatian Pelican
de: Krauskopfpelikan
fr: Pélican frisé
es: Pelícano Ceñudo
ja: ニシハイイロペリカン
cn: 卷羽鹈鹕
Pelecanus crispus
Vulnerable.

www.avitopia.net/bird.en/?vid=675108

adult

Herons - *Ardeidae*

The family of Herons occurs on all continents and on many islands. The body length ranges from 28 nm to 140 cm. The wings are large, the tail is short. Legs, toes and neck are long, the latter has a characteristic S-shape. They feed mainly on fish, but also eat other small animals. They mostly breed in colonies. The food brought in is regurgitated in front of the chicks.

Eurasian Bittern
de: Rohrdommel
fr: Butor étoilé
es: Avetoro Común
ja: サンカノゴイ
cn: 大麻鳽
Botaurus stellaris

www.avitopia.net/bird.en/?vid=750103

adult

Little Bittern
de: Zwergdommel
fr: Blongios nain
es: Mirasol Pequeño
ja: ヒメヨシゴイ
cn: 小苇鳽
Ixobrychus minutus

♂ adult

Grey Heron
de: Graureiher
fr: Héron cendré
es: Garza Real
ja: アオサギ
cn: 苍鹭

Ardea cinerea

www.avitopia.net/bird.en/?vid=750702
www.avitopia.net/bird.en/?wid=750702

adult

Purple Heron
de: Purpurreiher
fr: Héron pourpré
es: Garza Imperial
ja: ムラサキサギ
cn: 草鹭

Ardea purpurea

www.avitopia.net/bird.en/?vid=750710

adult

Great Egret
de: Silberreiher
fr: Grande Aigrette
es: Garceta Grande
ja: ダイサギ
cn: 大白鹭

Ardea alba

www.avitopia.net/bird.en/?vid=750711
www.avitopia.net/bird.en/?wid=750711

adult

Little Egret
de: Seidenreiher
fr: Aigrette garzette
es: Garceta Común
ja: コサギ
cn: 白鹭

Egretta garzetta

www.avitopia.net/bird.en/?vid=750803

breeding

Cattle Egret
de: Kuhreiher
fr: Héron garde-boeufs
es: Garcilla Bueyera
ja: アマサギ
cn: 牛背鷺
Bubulcus ibis

www.avitopia.net/bird.en/?vid=750901

breeding

Squacco Heron
de: Rallenreiher
fr: Crabier chevelu
es: Garcilla Cangrejera
ja: カンムリサギ
cn: 白翅黄池鷺
Ardeola ralloides

breeding

Black-crowned Night Heron
de: Nachtreiher
fr: Bihoreau gris
es: Martinete Común
ja: ゴイサギ
cn: 夜鷺
Nycticorax nycticorax

www.avitopia.net/bird.en/?vid=751501

breeding

Ibises - *Threskiornithidae*

The family of Ibises occurs in all warm regions of the world. The birds are 50 cm to 110 cm high. They have long wings and a short tail. The toes are connected by small webs. The long beak is either curved downwards or broad and spatulate. Most species are quite gregarious. They fly powerfully and can sail with the neck stretched out. Their food is very varied. The Sacred Ibis was the revered symbol of the god Thoth in ancient Egypt, but is now extinct in this country.

Glossy Ibis
   de: Sichler
   fr: Ibis falcinelle
   es: Morito Común
   ja: ブロンズトキ
   cn: 彩鹮
Plegadis falcinellus

www.avitopia.net/bird.en/?vid=775201

adult

Eurasian Spoonbill
   de: Löffler
   fr: Spatule blanche
   es: Espátula Común
   ja: ヘラサギ
   cn: 白琵鹭
Platalea leucorodia

www.avitopia.net/bird.en/?vid=776301

breeding

Ospreys - *Pandionidae*

There is only one species in the family of ospreys, but it is a true cosmopolitan: it occurs on all continents. The birds are around 60 cm tall with long wings and short tails. The beak is hook-shaped and the feet have warty sole pads. Fischdler are pure fish hunters who hover first and then plunges into the water, often submerging themselves completely. The fish are grasped with the claws and carried to a sitting site or a nest. The female breeds and cares for the young birds alone.

AU	Osprey
	de: Fischadler
	fr: Balbuzard pêcheur
	es: Águila Pescadora
	ja: ミサゴ
	cn: 鹗
	Pandion haliaetus

♀ adult

Birds of Prey - *Accipitridae*

The family of Birds of Prey is found worldwide with the exception of the Arctic, Antarctic and most of the oceanic islands. Birds of Prey are of various sizes (20 - 115 cm), they have long, round wings, curved claws and a short hooked bill. The sexes are almost the same, the female is almost always larger. All species are good fliers, and many sail well too. They mainly hunt live animals, only the vultures are scavengers. Even fishing species are among them.

AU	Bearded Vulture
	de: Bartgeier
	fr: Gypaète barbu
	es: Quebrantahuesos
	ja: ヒゲワシ
	cn: 胡兀鹫
	Gypaetus barbatus
	Near threatened.

www.avitopia.net/bird.en/?vid=875701
www.avitopia.net/bird.en/?wid=875701

adult

Egyptian Vulture
 de: Schmutzgeier
 fr: Vautour percnoptère
 es: Alimoche Común
 ja: エジプトハゲワシ
 cn: 白兀鹫
Neophron percnopterus
Endangered.

www.avitopia.net/bird.en/?vid=875801

adult

European Honey Buzzard
 de: Wespenbussard
 fr: Bondrée apivore
 es: Abejero Europeo
 ja: ヨアロッパハチクマ
 cn: 鹃头蜂鹰
Pernis apivorus

adult

Cinereous Vulture
 de: Mönchsgeier
 fr: Vautour moine
 es: Buitre Negro
 ja: クロハゲワシ
 cn: 秃鹫
Aegypius monachus
Near threatened.

www.avitopia.net/bird.en/?vid=876901

adult

Griffon Vulture
 de: Gänsegeier
 fr: Vautour fauve
 es: Buitre Leonado
 ja: ケアブシロエリハゲワシ
 cn: 西域兀赞
Gyps fulvus

www.avitopia.net/bird.en/?kom=877207
www.avitopia.net/bird.en/?vid=877207

adult

Short-toed Snake Eagle
  de: Schlangenadler
  fr: Circaète Jean-le-Blanc
  es: Culebrera Europea
  ja: チュウヒワシ
  cn: 短趾雕
Circaetus gallicus

adult — Drawing J.G.Keulemans

Lesser Spotted Eagle
  de: Schreiadler
  fr: Aigle pomarin
  es: Águila Pomerana
  ja: アシナガワシ
  cn: 小乌雕
Clanga pomarina

adult — Drawing E.Neale

Greater Spotted Eagle
  de: Schelladler
  fr: Aigle criard
  es: Águila Moteada
  ja: カラフトワシ
  cn: 乌雕
Clanga clanga
Vulnerable.

adult — Photo J.M.Garg

Booted Eagle
  de: Zwergadler
  fr: Aigle botté
  es: Águila calzada
  ja: ヒメクマタカ
  cn: 靴隼雕
Hieraaetus pennatus

adult — Photo W.J.Daunicht

Asian Imperial Eagle
  de:Kaiseradler
  fr: Aigle impérial
  es: Águila Imperial Oriental
  ja: カタジロワシ
  cn: 白肩雕
Aquila heliaca
Vulnerable.

adult

Golden Eagle
  de:Steinadler
  fr: Aigle royal
  es: Águila Real
  ja: イヌワシ
  cn: 金雕
Aquila chrysaetos

www.avitopia.net/bird.en/?vid=879106
www.avitopia.net/bird.en/?aud=879106

adult

Western Marsh Harrier
  de:Rohrweihe
  fr: Busard des roseaux
  es: Aguilucho Lagunero Occidental
  ja: ヨアロッパチュウヒ
  cn: 白头鹞
Circus aeruginosus

www.avitopia.net/bird.en/?vid=880101

♂ adult

Northern Harrier
  de:Kornweihe
  fr: Busard Saint-Martin
  es: Aguilucho Pálido
  ja: ハイイロチュウヒ
  cn: 白尾鹞
Circus cyaneus

www.avitopia.net/bird.en/?vid=880110

♂ adult

♂ adult

S4.0 **Pallid Harrier**
de: Steppenweihe
fr: Busard pâle
es: Aguilucho Papialbo
ja: ウスハイイロチュウヒ
cn: 草原鹞
Circus macrourus
Near threatened.

Photo Roger Clarke

♂ adult

AU **Montagu's Harrier**
de: Wiesenweihe
fr: Busard cendré
es: Aguilucho Cenizo
ja: ヒメハイイロチュウヒ
cn: 乌灰鹞
Circus pygargus

Photo W.J.Daunicht

♂ adult

PD **Levant Sparrowhawk**
de: Kurzfangsperber
fr: Épervier à pieds courts
es: Gavilán Griego
ja: レバントハイタカ
cn: 东雀鹰
Accipiter brevipes

Drawing J.G.Keulemans

♀ adult

AU **Eurasian Sparrowhawk**
de: Sperber
fr: Épervier d'Europe
es: Gavilán Común
ja: ハイタカ
cn: 雀鹰
Accipiter nisus

Photo W.J.Daunicht

Northern Goshawk
 de: Habicht
 fr: Autour des palombes
 es: Azor Común
 ja: オオタカ
 cn: 苍鹰
Accipiter gentilis

adult

Red Kite
 de: Rotmilan
 fr: Milan royal
 es: Milano Real
 ja: アカトビ
 cn: 赤鸢
Milvus milvus
Near threatened.

www.avitopia.net/bird.en/?vid=880601

adult

Black Kite
 de: Schwarzmilan
 fr: Milan noir
 es: Milano Negro
 ja: トビ
 cn: 黑鸢
Milvus migrans

www.avitopia.net/bird.en/?vid=880602

adult

White-tailed Eagle
 de: Seeadler
 fr: Pygargue à queue blanche
 es: Pigargo Europeo
 ja: オジロワシ
 cn: 白尾海雕
Haliaeetus albicilla

www.avitopia.net/bird.en/?vid=880802

adult

Rough-legged Buzzard
de: Raufußbussard
fr: Buse pattue
es: Busardo Calzado
ja: ケアシノスリ
cn: 毛脚鵟
Buteo lagopus

adult

Common Buzzard
de: Mäusebussard
fr: Buse variable
es: Busardo Ratonero
ja: ノスリ
cn: 普通鵟
Buteo buteo

www.avitopia.net/bird.en/?vid=881816
www.avitopia.net/bird.en/?aud=881816

adult

Long-legged Buzzard
de: Adlerbussard
fr: Buse féroce
es: Busardo Moro
ja: ニシオオノスリ
cn: 棕尾鵟
Buteo rufinus

adult

Bustards - *Otididae*

The family of Bustards occurs in Africa, Eurasia and Australia. The height of the birds ranges from 38 cm to 130 cm. They have long broad wings, long strong legs with three short toes and a long neck. The beak is short, strong and flattened. Their habitat are free plains. They fly well, but are downright ground birds. All species are omnivorous. In courtship, the throat pouch serves to amplify the voice. The nest is an unpadded recess in the ground in which 1 to 5 eggs are laid. The young birds flee the nest.

Great Bustard
  de: Großtrappe
  fr: Grande Outarde
  es: Avutarda Euroasiática
  ja: ノガン
  cn: 大鸨
Otis tarda
Vulnerable.

www.avitopia.net/bird.en/?vid=900101
www.avitopia.net/bird.en/?wid=900101

♂ adult

Little Bustard
  de: Zwergtrappe
  fr: Outarde canepetière
  es: Sisón Común
  ja: ヒメノガン
  cn: 小鸨
Tetrax tetrax
Near threatened.

♂ adult

Rails, Waterhens, and Coots - *Rallidae*

The family of Rails and Coots occurs worldwide except in the polar regions. Rails are at most medium-sized birds (15 - 50 cm) with short wings and very short tails. The toes are long and have swimming lobes in the coots. The sexes usually look the same. Almost all species swim well. Many only become active at dusk or are night birds. Some are able to fly long distances, while island species are partially flightless.

Corn Crake
de: Wachtelkönig
fr: Râle des genêts
es: Guión de Codornices
ja: ウズラクイナ
cn: 长脚秧鸡
Crex crex

adult
Photo Richard Wesley

Water Rail
de: Wasserralle
fr: Râle d'eau
es: Rascón Europeo
ja: クイナ
cn: 普通秧鸡
Rallus aquaticus

adult
Photo Pierre Dalous

Spotted Crake
de: Tüpfelsumpfhuhn
fr: Marouette ponctuée
es: Polluela Pintoja
ja: コモンクイナ
cn: 斑胸田鸡
Porzana porzana

adult
Photo Marek Szczepanek

Little Crake
  de: Kleines Sumpfhuhn
  fr: Marouette poussin
  es: Polluela Bastarda
  ja: コクイナ
  cn: 姬田鸡
Zapornia parva

♀ adult

Baillon's Crake
  de: Zwergsumpfhuhn
  fr: Marouette de Baillon
  es: Polluela Chica
  ja: ヒメクイナ
  cn: 小田鸡
Zapornia pusilla

adult

Common Moorhen
  de: Teichhuhn
  fr: Gallinule poule-d'eau
  es: Gallereta Común
  ja: バン
  cn: 黑水鸡
Gallinula chloropus

  www.avitopia.net/bird.en/?kom=1003303
  www.avitopia.net/bird.en/?vid=1003303

adult

Eurasian Coot
  de: Blässhuhn
  fr: Foulque macroule
  es: Focha Común
  ja: オオバン
  cn: 白骨顶
Fulica atra

  www.avitopia.net/bird.en/?vid=1003506
  www.avitopia.net/bird.en/?aud=1003506

adult

Cranes - *Gruidae*

The family of Cranes occurs worldwide except for South America. The body size ranges from 90 cm to 175 cm. They have long legs, necks, and beaks, but short tails. The inner arm feathers are converted into overhanging ornamental feathers. They owe their loud voice to an extension of the windpipe into the breastbone. They are excellent gliders who fly with their necks outstretched. They feed on a variety of animal and vegetable foods. They build their nests on the ground or even in water. The 2 or 3 eggs are incubated by both partners and both also look after the young birds who flee the nest.

Common Crane
de:Kranich
fr: Grue cendrée
es: Grulla Común
ja: クロヅル
cn:灰鹤
Grus grus

www.avitopia.net/bird.en/?kom=1125601
www.avitopia.net/bird.en/?vid=1125601

adult

Thick-knees - *Burhinidae*

The family of Thick-knees occurs on every continent on earth. They are 36 cm to 52 cm long. They have fairly long wings and a half-length tail. The legs are long and the feet have three webbed toes. The head is thick with large eyes. They are twilight and night birds and some species are gregarious. They rarely fly and feed on various small animals. Two eggs are usually laid in the nests on the ground. Both parents breed and later look after the young.

Eurasian Stone-curlew
de:Triel
fr: Oedicnème criard
es: Alcaraván Común
ja: イシチドリ
cn:石鸻
Burhinus oedicnemus

www.avitopia.net/bird.en/?vid=1200102

adult

Stilts and Avocets - *Recurvirostridae*

The family of Stilts and Avocets is widespread worldwide; the northern populations are migratory birds. The body length is 30 cm to 50 cm. They have very long legs and a thin beak that is straight or curved upwards. They fly well and can swim well. They search the mud in shallow waters for invertebrates. They nest in colonies, the nest-fleeing young birds are looked after by both parents. The defense of the offspring includes various distraction techniques from simulating a 'broken wing' to 'false brooding' in full view of a predator.

Black-winged Stilt
de: Stelzenläufer
fr: Échasse blanche
es: Cigüeñuela de Alas Negras
ja: セイタカシギ
cn: 黑翅长脚鹬
Himantopus himantopus

www.avitopia.net/bird.en/?vid=1250101
www.avitopia.net/bird.en/?aud=1250101

adult
Photo W.J.Daunicht

Pied Avocet
de: Säbelschnäbler
fr: Avocette élégante
es: Avoceta Común
ja: ソリハシセイタカシギ
cn: 反嘴鹬
Recurvirostra avosetta

www.avitopia.net/bird.en/?vid=1250301
www.avitopia.net/bird.en/?wid=1250301

adult
Photo W.J.Daunicht

Oystercatcher - *Haematopodidae*

The family of oystercatchers is found in temperate and tropical waters from Iceland and the Aleutian Islands to Cape Horn and Tasmania. The body length of the medium-sized birds is 32 cm to 45 cm. The legs are long and strong, the feet have small webs. The beak is long, strong and compressed at the sides. Their diet consists of mussels, crabs, worms and insects, but oysters are not the main ingredient. Outside of the breeding season, they are sociable and then gather in large flocks that can reach a few thousand. The chicks who flee the nest are looked after by both parents until they have fledged after five weeks.

Eurasian Oystercatcher
de: Austernfischer
fr: Huîtrier pie
es: Ostrero Euroasiático
ja: ミヤコドリ
cn: 蛎鹬
Haematopus ostralegus
Near threatened.

www.avitopia.net/bird.en/?vid=1300101
www.avitopia.net/bird.en/?wid=1300101

breeding

Plovers - *Charadriidae*

The plover family is global; many species are migratory birds. The body length ranges from 15 cm to 40 cm. Plover have a stocky body and long wings, the hind toe is receded or missing completely. They are ground birds that can run quickly, but also fly very well and quickly. In the vicinity of the nest or the young birds, the adult birds use seduction by simulating a broken wing and luring away a dangerous animal.

Grey Plover
de: Kiebitzregenpfeifer
fr: Pluvier argenté
es: Chorlito Gris
ja: ダイゼン
cn: 灰鸻
Pluvialis squatarola

breeding

European Golden Plover
 de: Goldregenpfeifer
 fr: Pluvier doré
 es: Chorlito Dorado Europeo
 ja: ヨアロッパムナグロ
 cn: 欧亚金鸻
Pluvialis apricaria

breeding

Northern Lapwing
 de: Kiebitz
 fr: Vanneau huppé
 es: Avefría Europea
 ja: タゲリ
 cn: 凤头麦鸡
Vanellus vanellus
Near threatened.

www.avitopia.net/bird.en/?vid=1325301
www.avitopia.net/bird.en/?aud=1325301

breeding

Kentish Plover
 de: Seeregenpfeifer
 fr: Pluvier à collier interrompu
 es: Chorlitejo Patinegro
 ja: シロチドリ
 cn: 环颈鸻
Charadrius alexandrinus

www.avitopia.net/bird.en/?vid=1325412

♂ non-breeding

Common Ringed Plover
 de: Sandregenpfeifer
 fr: Pluvier grand-gravelot
 es: Chorlitejo Grande
 ja: ハジロコチドリ
 cn: 剑鸻
Charadrius hiaticula

www.avitopia.net/bird.en/?vid=1325416

♂ breeding

Little Ringed Plover
de: Flussregenpfeifer
fr: Pluvier petit-gravelot
es: Chorlitejo Chico
ja: コチドリ
cn: 金眶鸻
Charadrius dubius

www.avitopia.net/bird.en/?vid=1325421

breeding

Eurasian Dotterel
de: Mornellregenpfeifer
fr: Pluvier guignard
es: Chorlito Carambolo
ja: コバシチドリ
cn: 小嘴鸻
Charadrius morinellus

adult

Sandpipers and Snipes - *Scolopacidae*

The family of Snipes is distributed worldwide, most of the species are migratory birds that sometimes cover great distances. The body length ranges from 13 cm to 60 cm. They usually live near water and outside the breeding season often form large flocks on the seashore. The diet consists of small invertebrates. The young birds leave the nest immediately after hatching.

Whimbrel
de: Regenbrachvogel
fr: Courlis corlieu
es: Zarapito Trinador
ja: チュウシャクシギ
cn: 中杓鹬
Numenius phaeopus

www.avitopia.net/bird.en/?vid=1450202

adult

Eurasian Curlew
 de:Großer Brachvogel
 fr: Courlis cendré
 es: Zarapito Real
 ja: ダイシャクシギ
 cn: 白腰杓鷸
Numenius arquata
Near threatened.

  www.avitopia.net/bird.en/?vid=1450208
  www.avitopia.net/bird.en/?aud=1450208

adult

Bar-tailed Godwit
 de: Pfuhlschnepfe
 fr: Barge rousse
 es: Aguja Colipinta
 ja: オオソリハシシギ
 cn: 斑尾塍鷸
Limosa lapponica
Near threatened.

  www.avitopia.net/bird.en/?vid=1450301

breeding

Black-tailed Godwit
 de: Uferschnepfe
 fr: Barge à queue noire
 es: Aguja Colinegra
 ja: オグロシギ
 cn: 黑尾塍鷸
Limosa limosa
Near threatened.

  www.avitopia.net/bird.en/?vid=1450302
  www.avitopia.net/bird.en/?aud=1450302

breeding

Ruddy Turnstone
 de: Steinwälzer
 fr: Tournepierre à collier
 es: Vuelvepiedras Común
 ja: キョウジョシギ
 cn: 翻石鷸
Arenaria interpres

breeding

Red Knot
　de:Knutt
　fr: Bécasseau maubèche
　es:Correlimos Gordo
　ja: コオバシギ
　cn:红腹滨鹬
Calidris canutus
Near threatened.

breeding

Ruff
　de:Kampfläufer
　fr: Combattant varié
　es:Combatiente
　ja: エリマキシギ
　cn:流苏鹬
Calidris pugnax

　www.avitopia.net/bird.en/?vid=1450604
　www.avitopia.net/bird.en/?wid=1450604

♂ breeding

Broad-billed Sandpiper
　de:Sumpfläufer
　fr: Bécasseau falcinelle
　es:Correlimos Falcinelo
　ja: キリアイ
　cn:阔嘴鹬
Calidris falcinellus

adult

Curlew Sandpiper
　de:Sichelstrandläufer
　fr: Bécasseau cocorli
　es:Correlimos Zarapitín
　ja: サルハマシギ
　cn:弯嘴滨鹬
Calidris ferruginea
Near threatened.

breeding

Temminck's Stint
de: Temminckstrandläufer
fr: Bécasseau de Temminck
es: Correlimos de Temminck
ja: オジロトウネン
cn: 青脚滨鹬
Calidris temminckii

breeding

Sanderling
de: Sanderling
fr: Bécasseau sanderling
es: Correlimos Tridáctilo
ja: ミユビシギ
cn: 三趾滨鹬
Calidris alba

non-breeding

Dunlin
de: Alpenstrandläufer
fr: Bécasseau variable
es: Correlimos Común
ja: ハマシギ
cn: 黑腹滨鹬
Calidris alpina

www.avitopia.net/bird.en/?vid=1450614

breeding

Little Stint
de: Zwergstrandläufer
fr: Bécasseau minute
es: Correlimos Menudo
ja: ニシトウネン
cn: 小滨鹬
Calidris minuta

www.avitopia.net/bird.en/?vid=1450618

breeding

Jack Snipe
de: Zwergschnepfe
fr: Bécassine sourde
es: Agachadiza Chica
ja: コシギ
cn: 姬鹬
Lymnocryptes minimus

adult

Eurasian Woodcock
de: Waldschnepfe
fr: Bécasse des bois
es: Chocha Perdiz
ja: ヤマシギ
cn: 丘鹬
Scolopax rusticola

adult

Great Snipe
de: Doppelschnepfe
fr: Bécassine double
es: Agachadiza Real
ja: ヨアロッパジシギ
cn: 斑腹沙锥
Gallinago media
Near threatened.

adult

Common Snipe
de: Bekassine
fr: Bécassine des marais
es: Agachadiza común
ja: タシギ
cn: 扇尾沙锥
Gallinago gallinago

adult

Red-necked Phalarope
de: Odinshühnchen
fr: Phalarope à bec étroit
es: Falaropo Picofino
ja: アカエリヒレアシシギ
cn: 红颈瓣蹼鹬
Phalaropus lobatus

♀ breeding

Red Phalarope
de: Thorshühnchen
fr: Phalarope à bec large
es: Falaropo Picogrueso
ja: ハイイロヒレアシシギ
cn: 灰瓣蹼鹬
Phalaropus fulicarius

♀ breeding

Common Sandpiper
de: Flussuferläufer
fr: Chevalier guignette
es: Andarríos Chico
ja: イソシギ
cn: 矶鹬
Actitis hypoleucos

www.avitopia.net/bird.en/?vid=1451401

adult

Green Sandpiper
de: Waldwasserläufer
fr: Chevalier cul-blanc
es: Andarríos Grande
ja: クサシギ
cn: 白腰草鹬
Tringa ochropus

www.avitopia.net/bird.en/?vid=1451501

adult

Spotted Redshank
de: Dunkler Wasserläufer
fr: Chevalier arlequin
es: Archibebe Oscuro
ja: ツルシギ
cn: 鹤鹬
Tringa erythropus

breeding

Common Greenshank
de: Grünschenkel
fr: Chevalier aboyeur
es: Archibebe Claro
ja: アオアシシギ
cn: 青脚鹬
Tringa nebularia

breeding

Marsh Sandpiper
de: Teichwasserläufer
fr: Chevalier stagnatile
es: Archibebe Fino
ja: コアオアシシギ
cn: 泽鹬
Tringa stagnatilis

breeding

Wood Sandpiper
de: Bruchwasserläufer
fr: Chevalier sylvain
es: Andarríos Bastardo
ja: タカブシギ
cn: 林鹬
Tringa glareola

www.avitopia.net/bird.en/?vid=1451512

adult

Common Redshank
  de:Rotschenkel
  fr: Chevalier gambette
  es: Archibebe Común
  ja: アカアシシギ
  cn:红脚鹬
  Tringa totanus

  www.avitopia.net/bird.en/?vid=1451513
  www.avitopia.net/bird.en/?aud=1451513

breeding

Coursers and Pratincoles - *Glareolidae*

The family of Coursers and Pratincoles is common in the Old World. They are 15 cm to 25 cm long. The Coursers have long legs, short wide wings and a longer beak, while the Pratincoles have long pointed wings, a forked tail, medium-long legs and a short beak. Coursers are ground birds and good runners, Pratincoles are good fliers whose flight style is reminiscent of swallows. The nest is dug out on the ground; both parents breed and take care of the precocial chicks.

Collared Pratincole
  de:Rotflügel-Brachschwalbe
  fr: Glaréole à collier
  es: Canastera Común
  ja:ニシツバメチドリ
  cn:领燕鸻
  Glareola pratincola

  www.avitopia.net/bird.en/?vid=1525501
  www.avitopia.net/bird.en/?aud=1525501

adult

Jaegers - *Stercorariidae*

The family of Jaegers is native to the arctic and subarctic areas of the northern and southern hemispheres. They migrate very far and can spend indefinite time at sea. The body length ranges from 40 cm to 60 cm. Their feet are webbed and have strong claws. The beak is strong and has a curved tip. Skuas are very fast and agile fliers. They breed near bird colonies and are aggressive predators and parasites there. They chase other birds until they vomit their food.

Great Skua
de: Skua
fr: Grand Labbe
es: Págalo Grande
ja: オオトウゾクカモメ
cn: 北贼鸥
Stercorarius skua

adult

Pomarine Jaeger
de: Spatelraubmöwe
fr: Labbe pomarin
es: Págalo Pomarino
ja: トウゾクカモメ
cn: 中贼鸥
Stercorarius pomarinus

adult

Parasitic Jaeger
de: Schmarotzerraubmöwe
fr: Labbe parasite
es: Págalo Parásito
ja: クロトウゾクカモメ
cn: 短尾贼鸥
Stercorarius parasiticus

adult, dark phase

Long-tailed Jaeger
de:Falkenraubmöwe
fr: Labbe à longue queue
es: Págalo Rabero
ja: シロハラトウゾクカモメ
cn: 长尾贼鸥
Stercorarius longicaudus

adult

Auks - *Alcidae*

The family of Auks can be found throughout the Arctic region, the North Atlantic and the North Pacific. Their body length ranges from 17 cm to 70 cm. They are clumsy birds with short wings, feet and tails. Alks come ashore only to breed, all swim and dive well, but do not fly well. They mainly eat fish, which they chase with flapping wings underwater. They breed on ledges, in caves or niches.

Common Murre
de: Trottellumme
fr: Guillemot marmette
es: Arao Común
ja: ウミガラス
cn: 崖海鸦
Uria aalge

breeding

Gulls - *Laridae*

The family of Gulls is found worldwide, most of the species are migratory birds. The body length ranges from 20 cm to 75 cm. Gulls are strongly built, they have long, pointed wings and a rather long tail. Their feet are webbed. They are very good fliers who often sail or glide. They can also swim well, but few species dive. They often breed in large colonies.

Black-legged Kittiwake
- de: Dreizehenmöwe
- fr: Mouette tridactyle
- es: Gaviota Tridáctila
- ja: ミツユビカモメ
- cn: 三趾鸥

Rissa tridactyla

breeding

Slender-billed Gull
- de: Dünnschnabelmöwe
- fr: Goéland railleur
- es: Gaviota Picofina
- ja: ハシボソカモメ
- cn: 细嘴鸥

Chroicocephalus genei

adult

Common Black-headed Gull
- de: Lachmöwe
- fr: Mouette rieuse
- es: Gaviota Reidora
- ja: ユリカモメ
- cn: 红嘴鸥

Chroicocephalus ridibundus

www.avitopia.net/bird.en/?kom=1600610
www.avitopia.net/bird.en/?vid=1600610
www.avitopia.net/bird.en/?wid=1600610

breeding

Little Gull
  de:Zwergmöwe
  fr: Mouette pygmée
  es: Gaviota Enana
  ja: ヒメカモメ
  cn: 小鸥
Hydrocoloeus minutus

breeding

Mediterranean Gull
  de:Schwarzkopfmöwe
  fr: Mouette mélanocéphale
  es: Gaviota Cabecinegra
  ja: ニシズグロカモメ
  cn: 地中海鸥
Ichthyaetus melanocephalus

breeding

Audouin's Gull
  de:Korallenmöwe
  fr: Goéland d'Audouin
  es: Gaviota de Audouin
  ja: アカハシカモメ
  cn: 地中海鸥
Ichthyaetus audouinii

adult

Mew Gull
  de:Sturmmöwe
  fr: Goéland cendré
  es: Gaviota Cana
  ja: カモメ
  cn: 海鸥
Larus canus

www.avitopia.net/bird.en/?vid=1601106

breeding

Michahellis Gull
de: Mittelmeermöwe
fr: Goéland leucophée
es: Gaviota Patiamarilla
ja: キアシセグロカモメ
cn: 黃腳銀鷗
Larus michahellis

www.avitopia.net/bird.en/?vid=1601112

breeding

Caspian Gull
de: Steppenmöwe
fr: Goéland pontique
es: Gaviota del Caspio
ja: キアシセグロカモメ
cn: 黄腿鸥
Larus cachinnans

adult

Lesser Black-backed Gull
de: Heringsmöwe
fr: Goéland brun
es: Gaviota Sombría
ja: ニシセグロカモメ
cn: 小黑背银鸥
Larus fuscus

adult

Little Tern
de: Zwergseeschwalbe
fr: Sterne naine
es: Charrancito Común
ja: コアジサシ
cn: 白额燕鸥
Sternula albifrons

breeding

Gull-billed Tern
 de:Lachseeschwalbe
 fr: Sterne hansel
 es: Pagaza Piconegra
 ja: ハシブトアジサシ
 cn: 鸥嘴噪鸥
Gelochelidon nilotica

adult — Photo Charles Lam

Caspian Tern
 de: Raubseeschwalbe
 fr: Sterne caspienne
 es: Pagaza Piquirroja
 ja: オニアジサシ
 cn: 红嘴巨鸥
Hydroprogne caspia

breeding — Photo B.Harry

Black Tern
 de: Trauerseeschwalbe
 fr: Guifette noire
 es: Fumarel Común
 ja: ハシグロクロハラアジサシ
 cn: 黑浮鸥
Chlidonias niger

breeding — Photo W.J.Daunicht

White-winged Tern
 de: Weißflügel-Seeschwalbe
 fr: Guifette leucoptère
 es: Fumarel Aliblanco
 ja: ハジロクロハラアジサシ
 cn: 白翅浮鸥
Chlidonias leucopterus

breeding — Photo Frank Vassen

Whiskered Tern
de: Weißbart-Seeschwalbe
fr: Guifette moustac
es: Fumarel cariblanco
ja: クロハラアジサシ
cn: 须浮鸥
Chlidonias hybrida

breeding

Common Tern
de: Flussseeschwalbe
fr: Sterne pierregarin
es: Charrán Común
ja: アジサシ
cn: 普通燕鸥
Sterna hirundo

♂ breeding

Arctic Tern
de: Küstenseeschwalbe
fr: Sterne arctique
es: Charrán Artico
ja: キョクアジサシ
cn: 北极燕鸥
Sterna paradisaea

breeding

Sandwich Tern
de: Brandseeschwalbe
fr: Sterne caugek
es: Charrán Patinegro
ja: サンドイッチアジサシ
cn: 白嘴端凤头燕鸥
Thalasseus sandvicensis

www.avitopia.net/bird.en/?vid=1602203

breeding

Pigeons and Doves - *Columbidae*

The family of Pigeons and Doves is found all over the world except in the coldest regions. The body lengths range from 15 cm to 84 cm. They have medium-sized wings and often a long tail. The beak is rather short and not strong. The sexes are mostly the same. Their diet is predominantly vegetarian. The naked young birds are fed 'pigeon milk', a secretion that is formed in the parents' goiter.

Common Pigeon
de: Felsentaube
fr: Pigeon biset
es: Paloma Bravía
ja: カワラバト(ドバト)
cn: 原鸽
Columba livia

www.avitopia.net/bird.en/?vid=1650101
www.avitopia.net/bird.en/?aud=1650101

Stock Dove
de: Hohltaube
fr: Pigeon colombin
es: Paloma Zurita
ja: ヒメモリバト
cn: 欧鸽
Columba oenas

www.avitopia.net/bird.en/?kom=1650106
www.avitopia.net/bird.en/?vid=1650106

Common Wood Pigeon
de: Ringeltaube
fr: Pigeon ramier
es: Paloma Torcaz
ja: モリバト
cn: 斑尾林鸽
Columba palumbus

www.avitopia.net/bird.en/?kom=1650109
www.avitopia.net/bird.en/?vid=1650109

European Turtle Dove
de: Turteltaube
fr: Tourterelle des bois
es: Tórtola Europea
ja: コキジバト
cn: 欧斑鸠
Streptopelia turtur
Vulnerable.

www.avitopia.net/bird.en/?vid=1650501
www.avitopia.net/bird.en/?wid=1650501
www.avitopia.net/bird.en/?aud=1650501

Eurasian Collared Dove
de: Türkentaube
fr: Tourterelle turque
es: Tórtola Turca
ja: シラコバト
cn: 灰斑鸠
Streptopelia decaocto

www.avitopia.net/bird.en/?kom=1650507
www.avitopia.net/bird.en/?vid=1650507

Cuckoos - *Cuculidae*

The family of Cuckoos is found worldwide except in the coldest areas. Many species are migratory birds. Cuckoos often have slender bodies and very long tails. The body length ranges from 17 cm to 70 cm. Except for the Roadrunners, the legs are short, with two toes pointing forward and two pointing backwards. The sexes are mostly the same. They feed mainly on insects. Many species are pronounced brood parasites. After hatching, the young cuckoos regularly push the other nest siblings out of the nest. The non-parasitic species are able to build nests. With the exception of the Anis, cuckoos are loners.

Great Spotted Cuckoo
de: Häherkuckuck
fr: Coucou geai
es: Críalo Europeo
ja: マダラカンムリカッコウ
cn: 大斑凤头鹃
Clamator glandarius

Common Cuckoo
 de:Kuckuck
 fr: Coucou gris
 es: Cuco Común
 ja: カッコウ
 cn:大杜鹃
Cuculus canorus

www.avitopia.net/bird.en/?kom=1728210
www.avitopia.net/bird.en/?aud=1728210

adult

Barn owls - *Tytonidae*

The family of Barn Owls includes only a few species, but one of them is a true cosmopolitan. It occurs on all continents and only avoids the colder areas of the earth. The body length ranges from 23 cm to 55 cm. The legs are quite long, the central claw is comb-like. The head is endowed with a conspicuous veil and has a curved beak. Barn owls are nocturnal hunters that fly noiselessly near the ground. They can hear directionally and are able to locate their prey by hearing only.

Barn Owl
 de:Schleiereule
 fr: Effraie des clochers
 es: Lechuza Común
 ja: メンフクロウ
 cn:仓鸮
Tyto alba

www.avitopia.net/bird.en/?vid=1750112

adult

Owls - *Strigidae*

The family of Owls is found worldwide. They have compact bodies (13 cm - 70 cm) and usually wide wings and rounded tails. The toes are strong, one of which is a turning toe that helps with grip. The head is large, the neck short. The eyes are directed rather forward, the beak is short and hook-shaped. Owls mainly hunt at night, benefiting from their noiseless flight and sharp hearing.

Eurasian Scops Owl
de: Zwergohreule
fr: Petit-duc scops
es: Autillo Europeo
ja: ヨアロッパコノハズク
cn: 西红角鸮
Otus scops

adult

Eurasian Eagle-Owl
de: Uhu
fr: Grand-duc d'Europe
es: Búho Real
ja: ワシミミズク
cn: 雕鸮
Bubo bubo

www.avitopia.net/bird.en/?vid=1776002
www.avitopia.net/bird.en/?aud=1776002

adult

Little Owl
de: Steinkauz
fr: Chevêche d'Athéna
es: Mochuelo Europeo
ja: コキンメフクロウ
cn: 纵纹腹小鸮
Athene noctua

www.avitopia.net/bird.en/?vid=1776704

adult

Tawny Owl
 de: Waldkauz
 fr: Chouette hulotte
 es: Cárabo Común
 ja: モリフクロウ
 cn: 灰林鸮
Strix aluco

www.avitopia.net/bird.en/?vid=1776904

adult

Ural Owl
 de: Habichtskauz
 fr: Chouette de l'Oural
 es: Cárabo Uralense
 ja: フクロウ
 cn: 长尾林鸮
Strix uralensis

www.avitopia.net/bird.en/?vid=1776914

♂ adult

Long-eared Owl
 de: Waldohreule
 fr: Hibou moyen-duc
 es: Búho Chico
 ja: トラフズク
 cn: 长耳鸮
Asio otus

www.avitopia.net/bird.en/?vid=1777001

adult

Short-eared Owl
 de: Sumpfohreule
 fr: Hibou des marais
 es: Búho Campestre
 ja: コミミズク
 cn: 短耳鸮
Asio flammeus

adult

Boreal Owl
de:Raufußkauz
fr: Nyctale de Tengmalm
es: Mochuelo Boreal
ja: キンメフクロウ
cn:鬼鸮
Aegolius funereus

adult

Nightjars - *Caprimulgidae*

The family of Nightjars is found all over the world, with the exception of the colder areas and New Zealand. The body length is between 19 cm and 30 cm, but elongated wing feathers appear in two species. Nightjars have long wings and tails, but their legs, toes, and claws are small. The beak is also small, but the throat is very wide. They are crepuscular and nocturnal and sit motionless on the ground or on a branch during the day. They feed on insects that they prey on in flight. The eggs are usually laid directly on the ground. There are also species among the nightjars that hibernate.

European Nightjar
de:Ziegenmelker
fr: Engoulevent d'Europe
es: Chotacabras Europeo
ja: ヨアロッパヨタカ
cn:欧亚夜鹰
Caprimulgus europaeus

adult

Swifts - *Apodidae*

The family of Swifts is globally distributed except in the coldest regions. They are small birds, 9 cm to 23 cm long. The wings are long and pointed, the legs and feet are very small and the beak is small with a crooked point and a wide throat. Sails are perfectly adapted to life in the air, and some species are able to spend the night in flight and to mate. They are excellent and fast fliers, on the other hand, many species cannot take off from the ground. The nests of a few species are made entirely of saliva and are considered a delicacy in Chinese cuisine. Some species of salangan have the exceptional echolocation capability. They use this to orient themselves in underground cave systems, where their nesting sites are.

Alpine Swift
de: Alpensegler
fr: Martinet à ventre blanc
es: Vencejo Real
ja: シロハラアマツバメ
cn: 高山雨燕
Apus melba

adult

Common Swift
de: Mauersegler
fr: Martinet noir
es: Vencejo Común
ja: ヨアロッパアマツバメ
cn: 雨燕
Apus apus

www.avitopia.net/bird.en/?kom=1926404
www.avitopia.net/bird.en/?vid=1926404
www.avitopia.net/bird.en/?aud=1926404

adult

Pallid Swift
de: Fahlsegler
fr: Martinet pâle
es: Vencejo Pálido
ja: ウルアマツバメ
cn: 苍雨燕
Apus pallidus

adult

Hoopoes - *Upupidae*

The family of Hoopoes is found in Eurasia and Africa and they are migratory birds. The body length is 16 cm to 32 cm. A hoopoe has broad round wings and a long squared tail. The third and fourth toes are fused at the root. The beak is long, slender and curved downwards. The feather crest can be erected and has black tips. They feed on insects and other invertebrates, but also on small vertebrates. They nest in a cave. The female breeds alone and is fed by the male.

Eurasian Hoopoe
de: Wiedehopf
fr: Huppe fasciée
es: Abubilla
ja: ヤツガシラ
cn: 戴胜
Upupa epops

www.avitopia.net/bird.en/?vid=2075101
www.avitopia.net/bird.en/?wid=2075101

adult

Photo W.J.Daunicht

Kingfishers - *Alcedinidae*

The family of Kingfishers are found worldwide except in the coldest areas and some islands. They are between 10 and 45 cm long. Kingfishers have a stocky body, short wings, and tiny to very long tails. The head is large, the neck short, and the beak long and thick. When hunting, they come down from a viewing point. Some species are capable of hovering flight. They breed in caves.

Common Kingfisher
de: Eisvogel
fr: Martin-pêcheur d'Europe
es: Martín Pescador Común
ja: カワセミ
cn: 普通翠鸟
Alcedo atthis

♂ adult

Photo Lukasz Lukasik

Bee-eaters - *Meropidae*

The family of bee-eaters occurs in tropical and subtropical areas on all continents of the Old World. The body length of the birds is 15 cm to 35 cm. The body is slender, the wings long and pointed, the tail is long, the legs and fused toes are slender. The long beak is compressed at the side and bent. The plumage is very different, with the bright colors red and green predominate. Most species are good fliers and snap their food, including bees and wasps, in the air. Many nest in colonies, both parents dig a tunnel up to one meter long in an embankment that ends in a small chamber. 2 to 5 eggs are laid there.

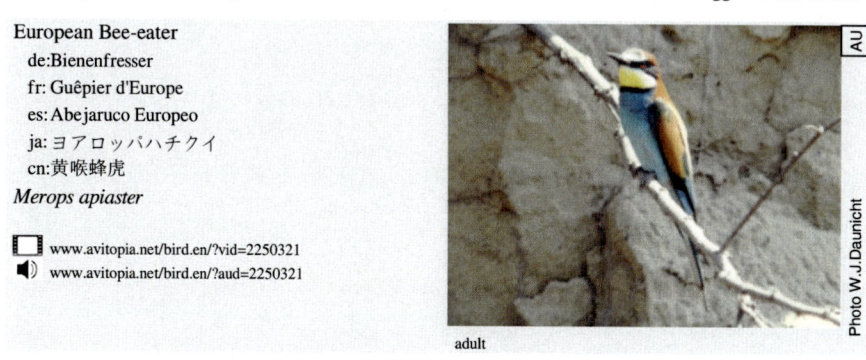

European Bee-eater
- de:Bienenfresser
- fr:Guêpier d'Europe
- es:Abejaruco Europeo
- ja:ヨアロッパハチクイ
- cn:黄喉蜂虎

Merops apiaster

www.avitopia.net/bird.en/?vid=2250321
www.avitopia.net/bird.en/?aud=2250321

adult

Photo W.J.Daunicht

Rollers - *Coraciidae*

The family of Rollers occurs in the warmer areas of the Old World, and in some cases they are downright long-distance migrants. The body length is 25 cm to 40 cm. They have long wings and a long tail that is often forked or notched. The beak is broad and hook-shaped. The plumage of most species is brightly colored. The voice is harsh, unmelodious and not very variable. They are good fliers, but like to hunt from a perch. Their courtship display in the air is spectacular and involves rolling. They nest in cavities.

European Roller
- de:Blauracke
- fr:Rollier d'Europe
- es:Carraca Europea
- ja:ニシブッポウソウ
- cn:蓝胸佛法僧

Coracias garrulus

www.avitopia.net/bird.en/?vid=2275101

adult

Photo W.J.Daunicht

Woodpeckers - *Picidae*

The family of Woodpeckers is found worldwide, but not in Madagascar, Australia and most of the Indonesian islands. The body length ranges from 9 cm to 60 cm. The feet have 3 or 4 toes, two of which point forward. They have a chunky head and a straight, usually powerful beak. The sexes are mostly different. Woodpeckers mostly inhabit trees. Most species feed on insects, which they pull out with their long and flexible tongue. When climbing vertical logs, the stiff tail serves as a support. They make a cave in a tree to breed.

Eurasian Wryneck
de: Wendehals
fr: Torcol fourmilier
es: Torcecuello Euroasiático
ja: アリスイ
cn: 蚁鴷
Jynx torquilla

adult

Lesser Spotted Woodpecker
de: Kleinspecht
fr: Pic épeichette
es: Pico Menor
ja: コアカゲラ
cn: 小斑啄木鸟
Dendrocopos minor

♂ adult

Middle Spotted Woodpecker
de: Mittelspecht
fr: Pic mar
es: Pico Mediano
ja: ヒメアカゲラ
cn: 中斑啄木鸟
Dendrocopos medius

www.avitopia.net/bird.en/?vid=2526118
www.avitopia.net/bird.en/?aud=2526118

♂ adult

White-backed Woodpecker
  de: Weißrückenspecht
  fr: Pic à dos blanc
  es: Pico Dorsiblanco
  ja: オオアカゲラ
  cn: 白背啄木鸟
Dendrocopos leucotos

♂ adult

Great Spotted Woodpecker
  de: Buntspecht
  fr: Pic épeiche
  es: Pico Picapinos
  ja: アカゲラ
  cn: 大斑啄木鸟
Dendrocopos major

www.avitopia.net/bird.en/?vid=2526120
www.avitopia.net/bird.en/?wid=2526120
www.avitopia.net/bird.en/?aud=2526120

♂ adult

Syrian Woodpecker
  de: Blutspecht
  fr: Pic syriaque
  es: Pico Sirio
  ja: カオジロアカゲラ
  cn: 叙利亚啄木鸟
Dendrocopos syriacus

♂ adult

Black Woodpecker
  de: Schwarzspecht
  fr: Pic noir
  es: Picamaderos Negro
  ja: クマゲラ
  cn: 黑啄木鸟
Dryocopus martius

www.avitopia.net/bird.en/?vid=2526706
www.avitopia.net/bird.en/?aud=2526706

♂ adult

European Green Woodpecker
de: Grünspecht
fr: Pic vert
es: Pito Real
ja: ヨアロッパアオゲラ
cn: 绿啄木鸟
Picus viridis

www.avitopia.net/bird.en/?vid=2526911
www.avitopia.net/bird.en/?aud=2526911

♀ adult

Falcons - *Falconidae*

The family of Falcons is found on every continent except Antarctica. Their length ranges from 15 cm to 65 cm. Hawks have long, pointed wings, a half-length tail and short legs that end in long toes with curved claws. The beak is short and usually has a so-called 'falcon tooth' in the upper beak. The flight is determined and fast. Some species strike their prey in flight after a chase, while other species take them on the ground after diving. In fact, the fastest fliers among birds belong to this family.

Lesser Kestrel
de: Rötelfalke
fr: Faucon crécerellette
es: Cernícalo Primilla
ja: ヒメチョウゲンボウ
cn: 黄爪隼
Falco naumanni

www.avitopia.net/bird.en/?kom=2576101

♂ adult

Common Kestrel
de: Turmfalke
fr: Faucon crécerelle
es: Cernícalo Vulgar
ja: チョウゲンボウ
cn: 红隼
Falco tinnunculus

www.avitopia.net/bird.en/?kom=2576102
www.avitopia.net/bird.en/?vid=2576102
www.avitopia.net/bird.en/?aud=2576102

♂ adult

Red-footed Falcon
 de: Rotfußfalke
 fr: Faucon kobez
 es: Cernícalo Patirrojo
 ja: ニシアカガシラチョウゲンボウ
 cn: 西红脚隼
Falco vespertinus
Near threatened.

♂ adult

Merlin
 de: Merlin
 fr: Faucon émerillon
 es: Esmerejón
 ja: コチョウゲンボウ
 cn: 灰背隼
Falco columbarius

♂ adult

Eurasian Hobby
 de: Baumfalke
 fr: Faucon hobereau
 es: Alcotán Europeo
 ja: チゴハヤブサ
 cn: 燕隼
Falco subbuteo

♂ adult

Saker Falcon
 de: Würgfalke
 fr: Faucon sacre
 es: Halcón Sacre
 ja: セアカアハヤブサ
 cn: 猎隼
Falco cherrug
Endangered.

adult

Peregrine Falcon
de: Wanderfalke
fr: Faucon pèlerin
es: Halcón Peregrino
ja: ハヤブサ
cn: 游隼
Falco peregrinus

adult

Shrikes - *Laniidae*

The family of Shrikes is found in North America, Eurasia, and Africa. Some species are downright long-distance migrants. The body length ranges from 14 cm to 50 cm. They are slender birds with a long, narrow tail. The head is large with a strong beak with a hook at the tip and tomial teeth on the sides. They are good fliers and very aggressive. From a viewing point they pounce on insects, small reptiles, birds and mammals. They are known to impale their prey on a thorn. The nest is a bowl made of twigs, grass and leaves in a bush or tree.

Red-backed Shrike
de: Neuntöter
fr: Pie-grièche écorcheur
es: Alcaudón de Dorso Rojo
ja: セアカモズ
cn: 红背伯劳
Lanius collurio

♂ adult

Great Grey Shrike
de: Raubwürger
fr: Pie-grièche grise
es: Alcaudón Norteño
ja: オオモズ
cn: 灰伯劳
Lanius excubitor

adult

Southern Grey Shrike
   de: Südraubwürger
   fr: Pie-grièche méridionale
   es: Alcaudón Real
   ja: ミナミオオモズ
   cn: 南灰伯劳
Lanius meridionalis

adult

Lesser Grey Shrike
   de: Schwarzstirnwürger
   fr: Pie-grièche à poitrine rose
   es: Alcaudón Chico
   ja: ヒメオオモズ
   cn: 黑额伯劳
Lanius minor

adult

Woodchat Shrike
   de: Rotkopfwürger
   fr: Pie-grièche à tête rousse
   es: Alcaudón Común
   ja: ズアカモズ
   cn: 林(即鸟)伯劳
Lanius senator

♂ adult

Orioles - *Oriolidae*

The family of Orioles is widespread across all continents of the Old World, with an emphasis on the tropics. Some of the species are migratory birds. They have long wings and a moderately long tail. The body size ranges from 18 cm to 30 cm. The plumage has strong colors, often yellow and black. The sexes are usually different. Some species in New Guinea belonging to the Pitohuis are able to store the poisonous alkaloid batrachotoxin in their plumage and skin in order to defend themselves against parasites and/or predators. Most species build their nests as a hanging bowl high in a tree top.

Eurasian Golden Oriole
de: Pirol
fr: Loriot d'Europe
es: Oropéndola Dorada Europea
ja: ニシコウライウグイス
cn: 金黄鹂
Oriolus oriolus

www.avitopia.net/bird.en/?vid=4050213
www.avitopia.net/bird.en/?wid=4050213
www.avitopia.net/bird.en/?aud=4050213

♂ adult

Ravens - *Corvidae*

The family of Ravens occurs worldwide with the exception of New Zealand and some islands. The body length is between 18 cm and 70 cm; so among them are the largest of all songbirds. Ravens have powerful bills and often hold the food with their feet when eating. They are curious and one of the most intelligent species in the entire bird world.

Eurasian Jay
de: Eichelhäher
fr: Geai des chênes
es: Arrandejo Común
ja: カケス
cn: 松鸦
Garrulus glandarius

www.avitopia.net/bird.en/?kom=4176101
www.avitopia.net/bird.en/?vid=4176101
www.avitopia.net/bird.en/?aud=4176101

adult

Eurasian Magpie
  de:Elster
  fr: Pie bavarde
  es: Urraca de Pico Negro
  ja: カササギ
  cn: 喜鹊
Pica pica

  www.avitopia.net/bird.en/?vid=4176801
  www.avitopia.net/bird.en/?aud=4176801

adult

Spotted Nutcracker
  de: Tannenhäher
  fr: Cassenoix moucheté
  es: Cascanueces Moteado
  ja: ホシガラス
  cn: 星鸦
Nucifraga caryocatactes

adult

Red-billed Chough
  de: Alpenkrähe
  fr: Crave à bec rouge
  es: Chova de Pico Rojo
  ja: ベニハシガラス
  cn: 红嘴山鸦
Pyrrhocorax pyrrhocorax

adult

Alpine Chough
  de: Alpendohle
  fr: Chocard à bec jaune
  es: Chova de Pico Amarillo
  ja: キバシガラス
  cn: 黄嘴山鸦
Pyrrhocorax graculus

adult

Western Jackdaw
 de: Dohle
 fr: Choucas des tours
 es: Grajilla Común
 ja: ニシコクマルガラス
 cn: 寒鸦
Corvus monedula

www.avitopia.net/bird.en/?vid=4177401
www.avitopia.net/bird.en/?aud=4177401

adult

Rook
 de: Saatkrähe
 fr: Corbeau freux
 es: Graja Común
 ja: ミヤマガラス
 cn: 秃鼻乌鸦
Corvus frugilegus

www.avitopia.net/bird.en/?vid=4177416
www.avitopia.net/bird.en/?aud=4177416

adult

Hooded Crow
 de: Nebelkrähe
 fr: Corneille mantelée
 es: Corneja Cenicienta
 ja: ハイイロガラス
 cn: 冠小嘴乌鸦
Corvus cornix

www.avitopia.net/bird.en/?vid=4177429
www.avitopia.net/bird.en/?aud=4177429

adult

Northern Raven
 de: Kolkrabe
 fr: Grand Corbeau
 es: Cuervo Común
 ja: ワタリガラス
 cn: 渡鸦
Corvus corax

www.avitopia.net/bird.en/?kom=4177444
www.avitopia.net/bird.en/?aud=4177444

adult

Bearded Reedlings - *Panuridae*

The family of Bearded Reedlings has only one member, which is restricted to Eurasia. The body size ranges from 14 cm to 15.5 cm. Bearded Reedlings live in reed beds, which they rarely leave. They feed on insects and spiders, and in winter they also eat seeds. The nest is a deep bowl near the water.

Bearded Reedling
- de: Bartmeise
- fr: Panure à moustaches
- es: Bigotudo
- ja: ヒゲガラ
- cn: 文须雀

Panurus biarmicus

www.avitopia.net/bird.en/?vid=4375101
www.avitopia.net/bird.en/?aud=4375101

♂ adult
Photo W.J.Daunicht

Larks - *Alaudidae*

The family of Larks occurs all over the world with the exception of New Zealand and many islands. Larks are rather small birds (12 cm - 23 cm). In most species, the claw of the hind toe is long and pointed. They love open terrain and advance into the hottest deserts. They look for food on the ground and eat insects, snails, seeds and buds. The fluid requirement is often met from food. Some species protect their nest hollow on the windward side with a little stone wall. The chicks are looked after by both parents; the nestling period is very short.

Horned Lark
- de: Ohrenlerche
- fr: Alouette hausse-col
- es: Alondra Cornuda
- ja: ハマヒバリ
- cn: 角百灵

Eremophila alpestris

adult
Photo T.Bowman

Greater Short-toed Lark
- de: Kurzzehenlerche
- fr: Alouette calandrelle
- es: Terrera Común
- ja: ヒメコウテンシ
- cn: 大短趾百灵

Calandrella brachydactyla

adult
Photo J.M.Garg

Calandra Lark
  de:Kalanderlerche
  fr: Alouette calandre
  es: Calandria Común
  ja: クロコウテンシ
  cn: 草原百灵
  Melanocorypha calandra

Woodlark
  de:Heidelerche
  fr: Alouette lulu
  es: Alondra Totovía
  ja: モリヒバリ
  cn: 林百灵
  Lullula arborea

Eurasian Skylark
  de:Feldlerche
  fr: Alouette des champs
  es: Alondra Común
  ja: ニシヒバリ
  cn: 云雀
  Alauda arvensis

  www.avitopia.net/bird.en/?kom=4427002
  www.avitopia.net/bird.en/?aud=4427002

Crested Lark
  de:Haubenlerche
  fr: Cochevis huppé
  es: Cogujada Común
  ja: カンムリヒバリ
  cn: 凤头百灵
  Galerida cristata

Swallows - *Hirundinidae*

The family of Swallows is found all over the world with the exception of the coldest regions, many species are migratory birds. They are quite small birds with a body length of 10 cm to 23 cm. The wings are long and pointed, the legs and feet small, the beak short with a wide throat. Swallows are fast and extremely agile fliers. They feed exclusively on flying insects. They nest in dug earth caves, natural caves or bowl-shaped mud nests. They breed up to three times a year.

Sand Martin
de: Uferschwalbe
fr: Hirondelle de rivage
es: Avión Zapador
ja: ショウドウツバメ
cn: 崖沙燕
Riparia riparia

www.avitopia.net/bird.en/?vid=4450904
www.avitopia.net/bird.en/?aud=4450904

adult — Photo Aiwok

Eurasian Crag Martin
de: Felsenschwalbe
fr: Hirondelle de rochers
es: Avión Roquero
ja: チャイロツバメ
cn: 岩燕
Ptyonoprogne rupestris

adult — Photo Martien Brand

Barn Swallow
de: Rauchschwalbe
fr: Hirondelle rustique
es: Golondrina Común
ja: ツバメ
cn: 家燕
Hirundo rustica

www.avitopia.net/bird.en/?kom=4451201
www.avitopia.net/bird.en/?vid=4451201

adult — Photo W.J.Daunicht

Red-rumped Swallow [AU]
de: Rötelschwalbe
fr: Hirondelle rousseline
es: Golondrina Dáurica
ja: コシアカツバメ
cn: 金腰燕

Cecropis daurica

Photo W.J.Daunicht

adult

Common House Martin [AU]
de: Mehlschwalbe
fr: Hirondelle de fenêtre
es: Avión Común
ja: ニシイワツバメ
cn: 毛脚燕

Delichon urbicum

🔊 www.avitopia.net/bird.en/?kom=4451501
▭ www.avitopia.net/bird.en/?vid=4451501

Photo W.J.Daunicht

adult

Tits - *Paridae*

The family of Tits are found on all continents of the world except South America and Australia. They are small birds with a body length of 8 cm to 21 cm. Tits usually live in trees and are rather bad fliers. Some species hide food in autumn, which they bring out again in winter. Up to 14 eggs can belong to a brood and are incubated by the female alone.

Coal Tit [A3.0]
de: Tannenmeise
fr: Mésange noire
es: Carbonero Garrapinos
ja: ヒガラ
cn: 煤山雀

Periparus ater

Photo Maren Winter

adult

European Crested Tit
 de: Haubenmeise
 fr: Mésange huppée
 es: Herrerillo Capuchino
 ja: カンムリガラ
 cn: 凤头山雀
Lophophanes cristatus

Sombre Tit
 de: Trauermeise
 fr: Mésange lugubre
 es: Carbonero Lúgubre
 ja: バルカンコガラ
 cn: 暗山雀
Poecile lugubris

Marsh Tit
 de: Sumpfmeise
 fr: Mésange nonnette
 es: Carbonero Palustre
 ja: ハシブトガラ
 cn: 沼泽山雀
Poecile palustris

Willow Tit
 de: Weidenmeise
 fr: Mésange boréale
 es: Carbonero Sibilino
 ja: コガラ
 cn: 褐头山雀
Poecile montanus

Blue Tit
de: Blaumeise
fr: Mésange bleue
es: Herrerillo Común
ja: アオガラ
cn: 青山雀

Cyanistes caeruleus

www.avitopia.net/bird.en/?kom=4500801

adult

Great Tit
de: Kohlmeise
fr: Mésange charbonnière
es: Carbonero Común
ja: シジュウカラ
cn: 大山雀

Parus major

www.avitopia.net/bird.en/?kom=4501102
www.avitopia.net/bird.en/?aud=4501102

♂ adult

Penduline Tits - *Remizidae*

The family of Penduline-Tits is common to North America, Eurasia and Africa. At 7.5 to 11 cm, they are very small birds with a pointed conical beak and relatively strong toes. Penduline-Tits build elaborate nests from plant wool with a side entrance that hang at the end of a branch and sway in the wind. An African species builds a conspicuous entrance that ends in a cul-de-sac, as well as an inconspicuous slit above it as a true entrance.

Eurasian Penduline Tit
de: Beutelmeise
fr: Rémiz penduline
es: Baloncito Común
ja: ニシツリスガラ
cn: 欧亚攀雀

Remiz pendulinus

♂ adult

Long-tailed Tits - *Aegithalidae*

The family of Long-tailed Tits is found in North America and Eurasia. They are tiny to small birds with a body length of 8.5 cm to 14 cm. They have short round wings and short bills. The tail is disproportionately long. Long-tailed tits build relatively tall closed nests with a side entrance near the top. The nests are camouflaged with lichen on the outside and lined with lots of small feathers on the inside. The female alone incubates a clutch of up to 12 eggs.

Long-tailed Bushtit
  de: Schwanzmeise
  fr: Mésange à longue queue
  es: Satrecito de Cola Larga
  ja: エナガ
  cn: 银喉长尾山雀
Aegithalos caudatus

  www.avitopia.net/bird.en/?vid=4550201

adult

Photo W.J.Daunicht

Nuthatches - *Sittidae*

The family of Nuthatches occurs in North America and Eurasia. They are rather small birds with a body length of 10.5 cm to 19.5 cm. The wings are long, the tail and legs are short, and the toes are relatively long. The beak is long, straight, pointed and quite powerful. Many species have a dark streak through the eyes. Nuthatches can also walk head down. They nest in caves in trees or rocks. The entrance is often plastered with mud. Or they build clay nests with a small entrance tube.

Eurasian Nuthatch
  de: Kleiber
  fr: Sittelle torchepot
  es: Sita de Eurasia
  ja: ゴジュウガラ
  cn: 普通鸸
Sitta europaea

  www.avitopia.net/bird.en/?kom=4575104
  www.avitopia.net/bird.en/?vid=4575104
  www.avitopia.net/bird.en/?aud=4575104

adult

Photo Smudge 9000

Western Rock Nuthatch
de: Felsenkleiber
fr: Sittelle de Neumayer
es: Sita Roquera
ja: イワゴジュウガラ
cn: 岩䳭
Sitta neumayer

adult

Wallcreepers - *Tichodromidae*

The family of Wallcreepers is common in Eurasia from Spain to eastern China. The body size is 15.5 to 17 cm. The plumage is mainly blue-gray, but when the wings are closed, the most noticeable features are largely hidden. When the wings are open, their extraordinary crimson-colored wing feathers are visible. Their habitat is in the mountains at altitudes from 1000 m to 3000 m and they only move to lower altitudes in winter. Wallcreepers are insectivores that feed on small invertebrates that they glean off the rock surface. They nest in crevices or caves that they pad out.

Wallcreeper
de: Mauerläufer
fr: Tichodrome échelette
es: Trepa Paredes
ja: カベバシリ
cn: 红翅旋壁雀
Tichodroma muraria

♀

Holarctic Treecreepers - *Certhiidae*

The family of Treecreepers occurs in North America, Eurasia, and Africa. The small birds (11 cm - 15 cm) have short wings, a long, slender and curved beak and very large feet with long claws. They use the short stiff tail as a support when climbing. They climb the trunks with jerky movements and fly from above to the base of the next tree. Their diet consists entirely of insects. The nest is built from natural materials as a bowl in a gap or behind protruding bark.

Eurasian Treecreeper
  de:Waldbaumläufer
  fr: Grimpereau des bois
  es: Agateador Norteño
  ja: キバシリ
  cn:旋木雀
Certhia familiaris

adult
Photo Frank Vassen

Short-toed Treecreeper
  de:Gartenbaumläufer
  fr: Grimpereau des jardins
  es: Agateador Común
  ja: タンシキバシリ
  cn:短趾旋木雀
Certhia brachydactyla

  www.avitopia.net/bird.en/?kom=4625105
  www.avitopia.net/bird.en/?vid=4625105

adult
Photo W.D.G.Daunicht

Wrens - *Troglodytidae*

The family of Wrens is found across America, Europe, and much of Asia. The small birds (10 cm to 22 cm) have short wings and strong legs and feet. The plumage is colored in different shades of brown with bands and spots. Most species fly little, but some can run quickly. They mainly eat insects, spiders and other invertebrates. They sing very lively and fiercely defend their territory.

Eurasian Wren
de: Zaunkönig
fr: Troglodyte mignon
es: Chochín Común
ja: ミソサザイ
cn: 鹪鹩

Troglodytes troglodytes

www.avitopia.net/bird.en/?kom=4650710
www.avitopia.net/bird.en/?aud=4650710
www.avitopia.net/bird.en/?eud=4650710

adult

Dippers - *Cinclidae*

The family od Dippers is found in America and Eurasia. The birds are clumsy with short wings and a short tail that is frequently cocked. The body length ranges from 14 cm to 23 cm. They have dense plumage and are well adapted to their habitat, fast-flowing mountain streams. They use their wings to swim underwater and can also walk on the bottom of the stream. They feed on invertebrates. The spherical nest is created in caves, often under a waterfall.

White-throated Dipper
de: Wasseramsel
fr: Cincle plongeur
es: Mirlo Acuático de Garganta Blanca
ja: ムナジロカワガラス
cn: 河乌

Cinclus cinclus

adult

Goldcrests/Kinglets - *Regulidae*

The family of Kinglets and Goldcrests is found in North America and Eurasia, the northern populations are migratory birds. They are very small birds with a lenght of 8 cm to 11 cm. They have medium-length wings and a thin, straight and pointed beak. They are particularly well adapted to the coniferous forest and eat almost exclusively insects. The suspended nest in the shape of a cup is very complex and can contain up to 2500 feathers.

Goldcrest
  de: Wintergoldhähnchen
  fr: Roitelet huppé
  es: Reyezuelo Sencillo
  ja: キクイタダキ
  cn: 戴菊
Regulus regulus

Firecrest
  de: Sommergoldhähnchen
  fr: Roitelet à triple bandeau
  es: Reyezuelo Listado
  ja: マミジロキクイタダキ
  cn: 火冠戴菊
Regulus ignicapilla

Bush-Warblers and allies - *Scotocercidae*

The family of Bush-warblers and allies occurs in Africa, Eurasia and some islands in the Pacific. The family was recently separated from the family of Old World Warblers based on DNA testing. They are small birds between 7 cm and 15 cm in length. Some species have very short tails. The plumage shows no bright colors.

Cetti's Warbler
 de: Seidensänger
 fr: Bouscarle de Cetti
 es: Ruiseñor Bastardo de Cetti
 ja: ヨアロッパウグイス
 cn: 宽尾树莺
Cettia cetti

adult

Leaf Warblers - *Phylloscopidae*

The warbler family of Leaf-warblers occurs in Eurasia and Africa and only one species reaches America, Alaska, but many species are long-distance migrants. With 9 cm to 14 cm, they are rather small birds that are predominantly greenish or brownish in color. They live in dense vegetation and feed on insects. The nests are created at a low height, are closed and have a side entrance.

Willow Warbler
 de: Fitis
 fr: Pouillot fitis
 es: Mosquitero de los Sauces
 ja: キタヤナギムシクイ
 cn: 欧柳莺
Phylloscopus trochilus

adult

Common Chiffchaff
 de: Zilpzalp
 fr: Pouillot véloce
 es: Mosquitero Común
 ja: チフチャフ
 cn: 叽喳柳莺
Phylloscopus collybita

www.avitopia.net/bird.en/?kom=4850108

♂ adult

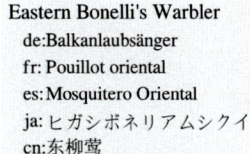

Eastern Bonelli's Warbler
de: Balkanlaubsänger
fr: Pouillot oriental
es: Mosquitero Oriental
ja: ヒガシボネリアムシクイ
cn: 东柳莺
Phylloscopus orientalis

Wood Warbler
de: Waldlaubsänger
fr: Pouillot siffleur
es: Mosquitero Silbador
ja: モリムシクイ
cn: 林柳莺
Phylloscopus sibilatrix

Reed-Warblers and allies - *Acrocephalidae*

The family of Reed-Warblers is native to Africa, Eurasia and Australia to Oceania. Many species are migratory birds. This family was only recently recognized based on DNA studies. The body size ranges from 11.5 cm to 18 cm. Almost all species have a relatively single-coloured plumage and can be better identified from their songs than from their appearance. However, the chants often contain imitations of other bird species. Most Reed-Warblers colonize wetlands or habitats near water.

Eastern Olivaceous Warbler
de: Blassspötter
fr: Hypolaïs pâle
es: Zarcero Pálido
ja: ハイイロウタムシクイ
cn: 草绿篱莺
Iduna pallida

Olive-tree Warbler
de: Olivenspötter
fr: Hypolaïs des oliviers
es: Zarcero Grande
ja: オリアブウタムシクイ
cn: 橄榄篱莺
Hippolais olivetorum

adult

Melodious Warbler
de: Orpheusspötter
fr: Hypolaïs polyglotte
es: Zarcero Común
ja: ウタムシクイ
cn: 歌篱莺
Hippolais polyglotta

adult

Icterine Warbler
de: Gelbspötter
fr: Hypolaïs ictérine
es: Zarcero Icterino
ja: キイロウタムシクイ
cn: 绿篱莺
Hippolais icterina

adult

Moustached Warbler
de: Mariskensänger
fr: Lusciniole à moustaches
es: Carricerín Real
ja: マミジロヨシキリ
cn: 须苇莺
Acrocephalus melanopogon

adult

Sedge Warbler
de: Schilfrohrsänger
fr: Phragmite des joncs
es: Carricerín Común
ja: スゲヨシキリ
cn: 蒲苇莺
Acrocephalus schoenobaenus

🔊 www.avitopia.net/bird.en/?kom=4875505
🔊 www.avitopia.net/bird.en/?aud=4875505

♂ adult

Marsh Warbler
de: Sumpfrohrsänger
fr: Rousserolle verderolle
es: Carricero Políglota
ja: ヌマヨシキリ
cn: 湿地苇莺
Acrocephalus palustris

adult

Eurasian Reed Warbler
de: Teichrohrsänger
fr: Rousserolle effarvatte
es: Carricero Común
ja: ヨアロッパヨシキリ
cn: 芦苇莺
Acrocephalus scirpaceus

🔊 www.avitopia.net/bird.en/?aud=4875512

adult

Great Reed Warbler
de: Drosselrohrsänger
fr: Rousserolle turdoïde
es: Carricerín Tordal
ja: オオヨシキリ
cn: 大苇莺
Acrocephalus arundinaceus

adult

Grassbirds and allies - *Locustellidae*

The family of grassbirds is common in Africa, Eurasia, and Australia. This family was only recently recognized on the basis of DNA studies. They are small, slender birds, the long tail is rounded and graduated. Grass warblers live in a wide variety of habitats, from grasslands to forests.

River Warbler
de: Schlagschwirl
fr: Locustelle fluviatile
es: Buscarla Fluvial
ja: カワセンニュウ
cn: 河蝗莺

Locustella fluviatilis

Savi's Warbler
de: Rohrschwirl
fr: Locustelle luscinioïde
es: Buscarla Unicolor
ja: ヌマセンニュウ
cn: 鸲蝗莺

Locustella luscinioides

🔊 www.avitopia.net/bird.en/?aud=4900609

Common Grasshopper Warbler
de: Feldschwirl
fr: Locustelle tachetée
es: Buscarla Pintoja
ja: ヤチセンニュウ
cn: 黑斑蝗莺

Locustella naevia

Cisticolas and allies - *Cisticolidae*

The family of Cisticolas and allies occurs in the Old World, but is confined to temperate and warm areas. This family was recently separated from the earlier extensive family of Old World Warblers and reorganized based on DNA studies. They are small birds between 9 cm and 20 cm in length. Many of the species have long and graduated tails. They live on small invertebrates. The nests are made with a lot of effort, some species, the tailor birds, literally sew large leaves together to give the nest a cover.

Zitting Cisticola
 de: Cistensänger
 fr: Cisticole des joncs
 es: Buitrón Común
 ja: セッカ
 cn: 棕扇尾莺
Cisticola juncidis

adult — Photo Anton Croos

Warblers - *Sylviidae*

The family of Old World Warblers is restricted to Eurasia and Africa, and many of the species are long-distance migrants. They are small birds with a body size of 11 cm to 15 cm. They feed on insects.

Eurasian Blackcap
 de: Mönchsgrasmücke
 fr: Fauvette à tête noire
 es: Curruca Capirotada
 ja: ズグロムシクイ
 cn: 黑顶林莺
Sylvia atricapilla

🔊 www.avitopia.net/bird.en/?kom=5000204

♂ adult — Photo W.J.Daunicht

Garden Warbler
 de: Gartengrasmücke
 fr: Fauvette des jardins
 es: Curruca Mosquitera
 ja: ニワムシクイ
 cn: 庭园林莺
Sylvia borin

♂ adult — Photo W.D.G.Daunicht

Barred Warbler
 de:Sperbergrasmücke
 fr: Fauvette épervière
 es: Curruca Gavilana
 ja: シマムシクイ
 cn: 横斑林莺
 Sylvia nisoria

adult

Lesser Whitethroat
 de:Klappergrasmücke
 fr: Fauvette babillarde
 es: Curruca Pequeña
 ja: コノドジロムシクイ
 cn: 白喉林莺
 Sylvia curruca

adult

Eastern Orphean Warbler
 de:Dickschnabel-Grasmücke
 fr: Fauvette à gros bec
 es: Curruca Mirlona Oriental
 ja: ヒガシメジロムシクイ
 cn: 东歌林莺
 Sylvia crassirostris

adult

Subalpine Warbler
 de:Weißbart-Grasmücke
 fr: Fauvette passerinette
 es: Curruca Carrasqueña
 ja: シラヒゲムシクイ
 cn: 亚高山林莺
 Sylvia cantillans

♂ adult

Sardinian Warbler
de: Samtkopf-Grasmücke
fr: Fauvette mélanocéphale
es: Curruca de Cabeza Negra
ja: クロガシラムシクイ
cn: 黑头林莺
Sylvia melanocephala

♂ adult

Common Whitethroat
de: Dorngrasmücke
fr: Fauvette grisette
es: Curruca Zarcera
ja: ノドジロムシクイ
cn: 灰白喉林莺
Sylvia communis

♂ adult

Flycatchers - *Muscicapidae*

The family of Flycatchers occurs in Africa, Europe and Asia. Many species are long-distance migrants. The body length ranges from 9 cm to 20 cm. Flycatchers have relatively long legs and are often strikingly colored. They mainly eat insects, but also plant-based foods.

Spotted Flycatcher
de: Grauschnäpper
fr: Gobemouche gris
es: Papamoscas Gris
ja: ムナフヒタキ
cn: 斑鶲
Muscicapa striata

adult

Rufous-tailed Scrub Robin
de: Heckensänger
fr: Agrobate roux
es: Alzacola
ja: オタテヤブコマドリ
cn: 棕薮鸲
Cercotrichas galactotes

adult

European Robin
de: Rotkehlchen
fr: Rougegorge familier
es: Petirrojo Europeo
ja: ヨアロッパコマドリ
cn: 欧亚鸲
Erithacus rubecula

www.avitopia.net/bird.en/?kom=5252001
www.avitopia.net/bird.en/?aud=5252001

adult

Thrush Nightingale
de: Sprosser
fr: Rossignol progné
es: Ruiseñor Ruso
ja: ヤブサヨナキドリ
cn: 欧歌鸲
Luscinia luscinia

adult

Common Nightingale
de: Nachtigall
fr: Rossignol philomèle
es: Ruiseñor
ja: サヨナキドリ
cn: 新疆歌鸲
Luscinia megarhynchos

www.avitopia.net/bird.en/?aud=5253202

adult

Bluethroat
  de:Blaukehlchen
  fr: Gorgebleue à miroir
  es: Pechiazul
  ja: オガワコマドリ
  cn: 蓝喉歌鸲
Luscinia svecica

♂ adult

Red-breasted Flycatcher
  de:Zwergschnäpper
  fr: Gobemouche nain
  es: Papamoscas Papirrojo
  ja: オジロビタキ
  cn: 红喉姬鹟
Ficedula parva

♂ adult

Semicollared Flycatcher
  de:Halbringschnäpper
  fr: Gobemouche à demi-collier
  es: Papamoscas Semicollarino
  ja: ハンエリヒタキ
  cn: 半领姬鹟
Ficedula semitorquata

♂ breeding

European Pied Flycatcher
  de:Trauerschnäpper
  fr: Gobemouche noir
  es: Papamoscas Cerrojillo
  ja: マダラヒタキ
  cn: 斑姬鹟
Ficedula hypoleuca

♂ adult

Collared Flycatcher
de: Halsbandschnäpper
fr: Gobemouche à collier
es: Papamoscas Collarino
ja: シロエリヒタキ
cn: 白领姬鹟
Ficedula albicollis

♂ adult

Common Redstart
de: Gartenrotschwanz
fr: Rougequeue à front blanc
es: Colirrojo Real
ja: シロビタイジョウビタキ
cn: 红尾鸲
Phoenicurus phoenicurus

www.avitopia.net/bird.en/?kom=5254109

♂ adult

Black Redstart
de: Hausrotschwanz
fr: Rougequeue noir
es: Colirrojo Tizón
ja: クロジョウビタキ
cn: 赭红尾鸲
Phoenicurus ochruros

www.avitopia.net/bird.en/?vid=5254113
www.avitopia.net/bird.en/?aud=5254113

♂ adult

Rufous-tailed Rock Thrush
de: Steinrötel
fr: Monticole merle-de-roche
es: Roquero Rojo
ja: コシジロイソヒヨドリ
cn: 白背矶鸫
Monticola saxatilis

♂ breeding

Blue Rock Thrush
 de: Blaumerle
 fr: Monticole merle-bleu
 es: Roquero Solitario
 ja: イソヒヨドリ
 cn: 蓝矶鸫
Monticola solitarius

♂ adult

Whinchat
 de: Braunkehlchen
 fr: Tarier des prés
 es: Tarabilla Norteña
 ja: マミジロノビタキ
 cn: 草原石(即鸟)
Saxicola rubetra

♂ adult

European Stonechat
 de: Schwarzkehlchen
 fr: Tarier pâtre
 es: Tarabilla común
 ja: ニシノビタキ
 cn: 黑喉石(即鸟)
Saxicola rubicola

🔊 www.avitopia.net/bird.en/?kom=5254305
▶ www.avitopia.net/bird.en/?vid=5254305

♂ breeding

Northern Wheatear
 de: Steinschmätzer
 fr: Traquet motteux
 es: Collalba Gris
 ja: ハシグロヒタキ
 cn: 穗(即鸟)
Oenanthe oenanthe

♂ adult

Black-eared Wheatear
>   de: Mittelmeer-Steinschmätzer
>   fr: Traquet oreillard
>   es: Collalba Rubia
>   ja: カオグロサバクヒタキ
>   cn: 白顶(即鸟)
>   *Oenanthe hispanica*

♂ adult

Isabelline Wheatear
>   de: Isabellsteinschmätzer
>   fr: Traquet isabelle
>   es: Collalba Isabel
>   ja: イナバヒタキ
>   cn: 沙(即鸟)
>   *Oenanthe isabellina*

adult

Thrushes - *Turdidae*

The family of Thrushes is distributed worldwide and is even found on many small islands in the Pacific, only missing in Antarctica and New Zealand. But Blackbirds and Song thrushes have been introduced there and have reproduced so much that they are now among the most common birds. Thrushes mainly feed on insects and other invertebrates, but berries also play a role in winter.

Ring Ouzel
>   de: Ringdrossel
>   fr: Merle à plastron
>   es: Mirlo de Capa Blanca
>   ja: クビワツグミ
>   cn: 环颈鸫
>   *Turdus torquatus*

♂ adult

Common Blackbird
  de:Amsel
  fr: Merle noir
  es: Mirlo Común
  ja: クロウタドリ
  cn: 乌鸫

Turdus merula

  www.avitopia.net/bird.en/?vid=5276321
  www.avitopia.net/bird.en/?aud=5276321

♂ adult

Fieldfare
  de:Wacholderdrossel
  fr: Grive litorne
  es: Zorzal Real
  ja: ノハラツグミ
  cn: 田鸫

Turdus pilaris

  www.avitopia.net/bird.en/?kom=5276337
  www.avitopia.net/bird.en/?vid=5276337
  www.avitopia.net/bird.en/?aud=5276337

adult

Redwing
  de:Rotdrossel
  fr: Grive mauvis
  es: Zorzal de Alas Rojas
  ja: ワキアカツグミ
  cn: 白眉歌鸫

Turdus iliacus
Near threatened.

adult

Song Thrush
  de:Singdrossel
  fr: Grive musicienne
  es: Zorzal Común
  ja: ウタツグミ
  cn: 欧歌鸫

Turdus philomelos

  www.avitopia.net/bird.en/?kom=5276339
  www.avitopia.net/bird.en/?vid=5276339
  www.avitopia.net/bird.en/?aud=5276339

adult

Mistle Thrush
de:Misteldrossel
fr: Grive draine
es:Zorzal Charlo
ja:ヤドリギツグミ
cn:槲鸫
Turdus viscivorus

adult

Starlings - *Sturnidae*

The family of Starlings was originally only distributed in the Old World, but the common star was introduced in America and is now widespread there. The body length ranges from 18 cm to 43 cm. Many species have iridescent plumage. The tail is usually short, more rarely long. Unlike Thrushes, Starlings do not hop, but run with alternating steps. They fly well and the formation flights of large flocks of Starlings are impressive. Most species breed in tree hollows, but other nesting techniques also occur, including large community nests. Starlings are omnivores, one reason for their assertiveness as colonists.

Common Starling
de:Star
fr: Étourneau sansonnet
es:Estornino Pinto
ja:ホシムクドリ
cn:紫翅椋鸟
Sturnus vulgaris

www.avitopia.net/bird.en/?kom=5326201

♂ breeding

Accentors - *Prunellidae*

The range of the family of Accentors is restricted to Eurasia. They are small birds, 14.5 cm to 18 cm long. Most of the species live in the mountains, but one can also be found in urban parks. They are sedentary or withdraw from higher mountain areas in winter. The flexibility of the reproductive strategies is remarkable; all combinations of single and multiple males or females occur.

Alpine Accentor
de: Alpenbraunelle
fr: Accenteur alpin
es: Acentor Alpino
ja: イワヒバリ
cn: 领岩鹨
Prunella collaris

adult

Dunnock
de: Heckenbraunelle
fr: Accenteur mouchet
es: Acentor Común
ja: ヨアロッパカヤクグリ
cn: 林岩鹨
Prunella modularis

adult

Wagtails and Pipits - *Motacillidae*

The family of Wagtails is found worldwide except in the coldest areas, many species are migratory birds. They are slender birds with a body length of 13 cm to 22 cm. All species are ground birds, but they can also fly well. First and foremost, they are insectivores.

Western Yellow Wagtail
de: Wiesenschafstelze
fr: Bergeronnette printanière
es: Lavandera Boyera
ja: ツメナガセキレイ
cn: 黄鹡鸰
Motacilla flava

www.avitopia.net/bird.en/?vid=5475301

♂ adult

Grey Wagtail
de: Gebirgsstelze
fr: Bergeronnette des ruisseaux
es: Lavandera Cascadeña
ja: キセキレイ
cn: 灰鹡鸰
Motacilla cinerea

♂ adult

White Wagtail
de: Bachstelze
fr: Bergeronnette grise
es: Lavandera Blanca
ja: タイリクハクセキレイ
cn: 白鹡鸰
Motacilla alba

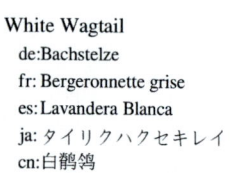

www.avitopia.net/bird.en/?kom=5475308
www.avitopia.net/bird.en/?vid=5475308

♂ adult

Tawny Pipit
de: Brachpieper
fr: Pipit rousseline
es: Bisbita Campestre
ja: ムジタヒバリ
cn: 平原鹨
Anthus campestris

adult

Meadow Pipit
de: Wiesenpieper
fr: Pipit farlouse
es: Bisbita Pratense
ja: マキバタヒバリ
cn: 草地鹨
Anthus pratensis
Near threatened.

♂ breeding

Tree Pipit
de:Baumpieper
fr: Pipit des arbres
es: Bisbita Arbóreo
ja: ヨアロッパビンズイ
cn:林鷚
Anthus trivialis

adult — Photo W.J.Daunicht

Red-throated Pipit
de:Rotkehlpieper
fr: Pipit à gorge rousse
es: Bisbita Gorgirrojo
ja: ムネアカタヒバリ
cn:红喉鷚
Anthus cervinus

adult — Photo Imran Shah

Water Pipit
de:Bergpieper
fr: Pipit spioncelle
es: Bisbita Alpino
ja: タヒバリ
cn:水鷚
Anthus spinoletta

adult — Photo Sandra

Waxwings - *Bombycillidae*

The family of Waxwings is native to the northern subarctic hemisphere. They are about 18 cm long. They are sociable tree dwellers who mainly feed on berries. In some years they invade areas that are outside the normal wintering range.

Bohemian Waxwing
de: Seidenschwanz
fr: Jaseur boréal
es: Ampelis Europeo
ja: キレンジャク
cn: 太平鸟
Bombycilla garrulus

♀ adult
Photo Tatiana Bulyonkova

Longspurs and Snow Buntings - *Calcariidae*

The family of Longspurs and Snow Buntings lives in the cooler regions of the northern hemisphere to beyond the Arctic Circle. In winter they move further south. Their body size ranges from 14 cm to 19 cm.

Snow Bunting
de: Schneeammer
fr: Bruant des neiges
es: Escribano Nival
ja: ユキホオジロ
cn: 雪鹀
Plectrophenax nivalis

♂ breeding
Photo T. Bowman

Old World Buntings - *Emberizidae*

The family od Old World Buntings is restricted zo the Old World, i.e. to Eurasia and Africa. They are stubby little birds with conical beaks. They feed on seeds. This family was only recently separated from the earlier, much larger-sized family of Buntings in the system of birds on the basis of DNA studies.

Yellowhammer
de: Goldammer
fr: Bruant jaune
es: Escribano Cerillo
ja: キアオジ
cn: 黄鹀
Emberiza citrinella

www.avitopia.net/bird.en/?kom=5850301
www.avitopia.net/bird.en/?vid=5850301
www.avitopia.net/bird.en/?aud=5850301

♂ adult

Pine Bunting
de: Fichtenammer
fr: Bruant à calotte blanche
es: Escribano Pinero
ja: シラガホオジロ
cn: 白头鹀
Emberiza leucocephalos

♂ adult

Cirl Bunting
de: Zaunammer
fr: Bruant zizi
es: Escribano Soteño
ja: ノドグロアオジ
cn: 黄道眉鹀
Emberiza cirlus

♂ adult

Rock Bunting
de: Zippammer
fr: Bruant fou
es: Escribano Montesino
ja: ヒゲホオジロ
cn: 灰眉岩鹀
Emberiza cia

♂ adult

Ortolan Bunting
de: Ortolan
fr: Bruant ortolan
es: Escribano Hortelano
ja: ズアオホオジロ
cn: 圃鹀
Emberiza hortulana

♂ adult

Cretzschmar's Bunting
de: Grauortolan
fr: Bruant cendrillard
es: Escribano Ceniciento
ja: ノドアカホオジロ
cn: 蓝头圃鹀
Emberiza caesia

♂ adult

Black-headed Bunting
de: Kappenammer
fr: Bruant mélanocéphale
es: Escribano de Cabeza Negra
ja: ズグロチャキンチョウ
cn: 黑头鹀
Emberiza melanocephala

♂ breeding

Common Reed Bunting
de: Rohrammer
fr: Bruant des roseaux
es: Escribano Palustre
ja: オオジュリン
cn: 芦鹀
Emberiza schoeniclus

www.avitopia.net/bird.en/?vid=5850341

♂ adult

Corn Bunting
de: Grauammer
fr: Bruant proyer
es: Escribano Triguero
ja: ハタホオジロ
cn: 黍鹀
Emberiza calandra

adult

Finches - *Fringillidae*

The family of Finches is widespread worldwide except for Australia and some oceanic islands. The body length is between 11 cm and 22 cm. They eat seeds and buds, insects almost only during the breeding season. The nest is built by the female from twigs, grass, moss and lichen in the form of a padded bowl.

Common Chaffinch
de: Buchfink
fr: Pinson des arbres
es: Pinzón Vulgar
ja: ズアオアトリ
cn: 苍头燕雀
Fringilla coelebs

www.avitopia.net/bird.en/?kom=6125101
www.avitopia.net/bird.en/?vid=6125101

♂ adult

Brambling
 de: Bergfink
 fr: Pinson du Nord
 es: Pinzón Real
 ja: アトリ
 cn: 燕雀
Fringilla montifringilla

www.avitopia.net/bird.en/?vid=6125103

Hawfinch
 de: Kernbeißer
 fr: Gros-bec casse-noyaux
 es: Pepitero Común
 ja: シメ
 cn: 锡嘴雀
Coccothraustes coccothraustes

www.avitopia.net/bird.en/?vid=6125503
www.avitopia.net/bird.en/?aud=6125503

Common Rosefinch
 de: Karmingimpel
 fr: Roselin cramoisi
 es: Camachuelo Carminoso
 ja: アカマシコ
 cn: 普通朱雀
Carpodacus erythrinus

Eurasian Bullfinch
 de: Gimpel
 fr: Bouvreuil pivoine
 es: Camachuelo Común
 ja: ウソ
 cn: 红腹灰雀
Pyrrhula pyrrhula

www.avitopia.net/bird.en/?aud=6127907

European Greenfinch
 de:Grünling
 fr: Verdier d'Europe
 es: Verderón Común
 ja: オオカワラヒワ
 cn:欧金翅雀
Chloris chloris

- www.avitopia.net/bird.en/?kom=6129001
- www.avitopia.net/bird.en/?vid=6129001
- www.avitopia.net/bird.en/?aud=6129001

♂ adult

Common Linnet
 de:Bluthänfling
 fr: Linotte mélodieuse
 es: Pardillo Común
 ja: ムネアカヒワ
 cn:赤胸朱顶雀
Linaria cannabina

- www.avitopia.net/bird.en/?vid=6129302
- www.avitopia.net/bird.en/?aud=6129302

♂ adult

Common Redpoll
 de:Birkenzeisig
 fr: Sizerin flammé
 es: Pardillo Sizerín
 ja: ベニヒワ
 cn:白腰朱顶雀
Acanthis flammea

adult

Lesser Redpoll
 de:Alpenbirkenzeisig
 fr: Sizerin cabaret
 es: Pardillo Alpino
 ja: イメベニヒワ
 cn:小朱顶雀
Acanthis cabaret

♂ breeding

Red Crossbill
  de: Fichtenkreuzschnabel
  fr: Bec-croisé des sapins
  es: Piquitureto Común
  ja: イスカ
  cn: 红交嘴雀
  Loxia curvirostra

♂ adult

European Goldfinch
  de: Stieglitz
  fr: Chardonneret élégant
  es: Jilguero Europeo
  ja: ゴシキヒワ
  cn: 红额金翅雀
  Carduelis carduelis

  www.avitopia.net/bird.en/?kom=6129701

adult

European Serin
  de: Girlitz
  fr: Serin cini
  es: Verdecillo
  ja: セリン
  cn: 欧洲丝雀
  Serinus serinus

  www.avitopia.net/bird.en/?vid=6129801
  www.avitopia.net/bird.en/?aud=6129801

♂ adult

Eurasian Siskin
  de: Erlenzeisig
  fr: Tarin des aulnes
  es: Lúgano
  ja: マヒワ
  cn: 黄雀
  Spinus spinus

  www.avitopia.net/bird.en/?vid=6129902

♂ adult

Sparrows - *Passeridae*

The sparrow family is native to Europe, Asia and Africa. However, one species managed to conquer the entire globe. The small birds are only 10 cm to 18 cm long. The conical beak indicates that they are grain eaters.

House Sparrow
  de: Haussperling
  fr: Moineau domestique
  es: Gorrión Doméstico
  ja: イエスズメ
  cn: 家麻雀
Passer domesticus

  www.avitopia.net/bird.en/?vid=6150202
  www.avitopia.net/bird.en/?aud=6150202

♂ adult

Spanish Sparrow
  de: Weidensperling
  fr: Moineau espagnol
  es: Gorrión Moruno
  ja: スペインスズメ
  cn: 黑胸麻雀
Passer hispaniolensis

♂ adult

Eurasian Tree Sparrow
  de: Feldsperling
  fr: Moineau friquet
  es: Gorrión Molinero
  ja: コガネスズメ
  cn: 麻雀
Passer montanus

  www.avitopia.net/bird.en/?kom=6150224

♂ adult

Rock Sparrow
   de: Steinsperling
   fr: Moineau soulcie
   es: Gorrión Chillón Rayado
   ja: イワスズメ
   cn: 石雀
Petronia petronia

adult

White-winged Snowfinch
   de: Schneefink
   fr: Niverolle alpine
   es: Gorrión Alpino de Alas Blancas
   ja: ユキスズメ
   cn: 白斑翅雪雀
Montifringilla nivalis

non-breeding

Indices of Names

Index of English Names

Accentor
  Alpine 103
Avocet
  Pied 41
Bean-Goose
  Taiga 8
Bee-eater
  European 67
Bittern
  Eurasian 26
  Little 26
Blackbird
  Common 101
Blackcap
  Eurasian 93
Bluethroat 97
Brambling 110
Bullfinch
  Eurasian 110
Bunting
  Black-headed 108
  Cirl 107
  Common Reed 109
  Corn 109
  Cretzschmar's 108
  Ortolan 108
  Pine 107
  Rock 108
  Snow 106
Bushtit
  Long-tailed 83
Bustard
  Great 37
  Little 37
Buzzard
  Common 36
  European Honey 31
  Long-legged 36
  Rough-legged 36
Capercaillie
  Western 16
Chaffinch
  Common 109

Chiffchaff
  Common 88
Chough
  Alpine 75
  Red-billed 75
Cisticola
  Zitting 93
Coot
  Eurasian 39
Cormorant
  Great 25
  Pygmy 24
Crake
  Baillon's 39
  Corn 38
  Little 39
  Spotted 38
Crane
  Common 40
Crossbill
  Red 112
Crow
  Hooded 76
Cuckoo
  Common 61
  Great Spotted 60
Curlew
  Eurasian 45
Dipper
  White-throated 86
Dotterel
  Eurasian 44
Dove
  Eurasian Collared 60
  European Turtle 60
  Stock 59
Duck
  Ferruginous 13
  Marbled 12
  Tufted 13
  White-headed 15
Dunlin 47
Dunnock 103

Eagle
  Asian Imperial 33
  Booted 32
  Golden 33
  Greater Spotted 32
  Lesser Spotted 32
  Short-toed Snake 32
  White-tailed 35
Eagle-Owl
  Eurasian 62
Egret
  Cattle 28
  Great 27
  Little 27
Falcon
  Peregrine 72
  Red-footed 71
  Saker 71
Fieldfare 101
Firecrest 87
Flamingo
  Greater 20
Flycatcher
  Collared 98
  European Pied 97
  Red-breasted 97
  Semicollared 97
  Spotted 95
Gadwall 11
Gannet
  Northern 24
Garganey 10
Godwit
  Bar-tailed 45
  Black-tailed 45
Goldcrest 87
Goldeneye
  Common 14
Goldfinch
  European 112
Goose
  Brant 9
  Greater White-fronted 8
  Greylag 8

Goshawk
  Northern 35
Grebe
  Black-necked 20
  Great Crested 20
  Horned 19
  Little 19
  Red-necked 19
Greenfinch
  European 111
Greenshank
  Common 50
Grouse
  Black 17
  Hazel 17
Gull
  Audouin's 55
  Caspian 56
  Common Black-headed 54
  Lesser Black-backed 56
  Little 55
  Mediterranean 55
  Mew 55
  Michahellis 56
  Slender-billed 54
Harrier
  Montagu's 34
  Northern 33
  Pallid 34
  Western Marsh 33
Hawfinch 110
Heron
  Black-crowned Night 28
  Grey 27
  Purple 27
  Squacco 28
Hobby
  Eurasian 71
Hoopoe
  Eurasian 66
Ibis
  Glossy 29
Jackdaw
  Western 76
Jaeger
  Long-tailed 53
  Parasitic 52
  Pomarine 52

Jay
  Eurasian 74
Kestrel
  Common 70
  Lesser 70
Kingfisher
  Common 66
Kite
  Black 35
  Red 35
Kittiwake
  Black-legged 54
Knot
  Red 46
Lapwing
  Northern 43
Lark
  Calandra 78
  Crested 78
  Greater Short-toed 77
  Horned 77
Linnet
  Common 111
Loon
  Black-throated 18
  Red-throated 18
Magpie
  Eurasian 75
Mallard 11
Martin
  Common House 80
  Eurasian Crag 79
  Sand 79
Merganser
  Common 14
  Red-breasted 15
Merlin 71
Moorhen
  Common 39
Murre
  Common 53
Nightingale
  Common 96
  Thrush 96
Nightjar
  European 64
Nutcracker
  Spotted 75
Nuthatch
  Eurasian 83

Western Rock 84
Oriole
  Eurasian Golden 74
Osprey 30
Ouzel
  Ring 100
Owl
  Barn 61
  Boreal 64
  Eurasian Scops 62
  Little 62
  Long-eared 63
  Short-eared 63
  Tawny 63
  Ural 63
Oystercatcher
  Eurasian 42
Partridge
  Grey 16
  Rock 16
Pelican
  Dalmatian 26
  Great White 25
Petrel
  European Storm 22
  Leach's Storm 22
  Wilson's Storm 22
Phalarope
  Red 49
  Red-necked 49
Pheasant
  Common 16
Pigeon
  Common 59
  Common Wood 59
Pintail
  Northern 11
Pipit
  Meadow 104
  Red-throated 105
  Tawny 104
  Tree 105
  Water 105
Plover
  Common Ringed 43
  European Golden 43
  Grey 42
  Kentish 43
  Little Ringed 44

Pochard
　Common 12
　Red-crested 12
Pratincole
　Collared 51
Ptarmigan
　Rock 17
Quail
　Common 15
Rail
　Water 38
Raven
　Northern 76
Redpoll
　Common 111
　Lesser 111
Redshank
　Common 51
　Spotted 50
Redstart
　Black 98
　Common 98
Redwing 101
Reedling
　Bearded 77
Robin
　European 96
　Rufous-tailed Scrub 96
Roller
　European 67
Rook 76
Rosefinch
　Common 110
Ruff 46
Sanderling 47
Sandpiper
　Broad-billed 46
　Common 49
　Curlew 46
　Green 49
　Marsh 50
　Wood 50
Scaup
　Greater 13
Scoter
　Black 14
　White-winged 13
Serin
　European 112

Shag
　European 25
Shearwater
　Cory's 21
　Manx 21
　Yelkouan 21
Shelduck
　Common 10
　Ruddy 10
Shoveler
　Northern 10
Shrike
　Great Grey 72
　Lesser Grey 73
　Red-backed 72
　Southern Grey 73
　Woodchat 73
Siskin
　Eurasian 112
Skua
　Great 52
Skylark
　Eurasian 78
Smew 14
Snipe
　Common 48
　Great 48
　Jack 48
Snowfinch
　White-winged 114
Sparrow
　Eurasian Tree 113
　House 113
　Rock 114
　Spanish 113
Sparrowhawk
　Eurasian 34
　Levant 34
Spoonbill
　Eurasian 29
Starling
　Common 102
Stilt
　Black-winged 41
Stint
　Little 47
　Temminck's 47
Stone-curlew
　Eurasian 40

Stonechat
　European 99
Stork
　Black 23
　White 23
Swallow
　Barn 79
　Red-rumped 80
Swan
　Mute 9
　Tundra 9
　Whooper 9
Swift
　Alpine 65
　Common 65
　Pallid 65
Teal
　Green-winged 12
Tern
　Arctic 58
　Black 57
　Caspian 57
　Common 58
　Gull-billed 57
　Little 56
　Sandwich 58
　Whiskered 58
　White-winged 57
Thrush
　Blue Rock 99
　Mistle 102
　Rufous-tailed Rock 98
　Song 101
Tit
　Blue 82
　Coal 80
　Eurasian Penduline 82
　European Crested 81
　Great 82
　Marsh 81
　Sombre 81
　Willow 81
Treecreeper
　Eurasian 85
　Short-toed 85
Turnstone
　Ruddy 45
Vulture
　Bearded 30
　Cinereous 31

Egyptian 31
Griffon 31
Wagtail
  Grey 104
  Western Yellow 103
  White 104
Wallcreeper 84
Warbler
  Barred 94
  Cetti's 88
  Common Grasshopper 92
  Eastern Bonelli's 89
  Eastern Olivaceous 89
  Eastern Orphean 94
  Eurasian Reed 91
  Garden 93
  Great Reed 91
  Icterine 90
  Marsh 91
  Melodious 90
  Moustached 90
  Olive-tree 90
  River 92
  Sardinian 95
  Savi's 92
  Sedge 91
  Subalpine 94
  Willow 88
  Wood 89
Waxwing
  Bohemian 106
Wheatear
  Black-eared 100
  Isabelline 100
  Northern 99
Whimbrel 44
Whinchat 99
Whitethroat
  Common 95
  Lesser 94
Wigeon
  Eurasian 11
Woodcock
  Eurasian 48
Woodlark 78
Woodpecker
  Black 69
  European Green 70
  Great Spotted 69
  Lesser Spotted 68

  Middle Spotted 68
  Syrian 69
  White-backed 69
Wren
  Eurasian 86
Wryneck
  Eurasian 68
Yellowhammer 107

Index of German Names

Adlerbussard 36
Alpenbirkenzeisig 111
Alpenbraunelle 103
Alpendohle 75
Alpenkrähe 75
Alpenschneehuhn 17
Alpensegler 65
Alpenstrandläufer 47
Amsel 101
Auerhuhn 16
Austernfischer 42
Bachstelze 104
Balkanlaubsänger 89
Bartgeier 30
Bartmeise 77
Basstölpel 24
Baumfalke 71
Baumpieper 105
Bekassine 48
Bergente 13
Bergfink 110
Bergpieper 105
Beutelmeise 82
Bienenfresser 67
Birkenzeisig 111
Birkhuhn 17
Blässgans 8
Blässhuhn 39
Blassspötter 89
Blaukehlchen 97
Blaumeise 82
Blaumerle 99
Blauracke 67
Bluthänfling 111
Blutspecht 69
Brachpieper 104
Brandgans 10
Brandseeschwalbe 58
Braunkehlchen 99
Bruchwasserläufer 50
Buchfink 109
Buntfuß-Sturmschwalbe 22
Buntspecht 69
Cistensänger 93
Dickschnabel-Grasmücke 94
Dohle 76

Doppelschnepfe 48
Dorngrasmücke 95
Dreizehenmöwe 54
Drosselrohrsänger 91
Dünnschnabelmöwe 54
Dunkler Wasserläufer 50
Eichelhäher 74
Eisvogel 66
Elster 75
Erlenzeisig 112
Fahlsegler 65
Falkenraubmöwe 53
Fasan 16
Feldlerche 78
Feldschwirl 92
Feldsperling 113
Felsenkleiber 84
Felsenschwalbe 79
Felsentaube 59
Fichtenammer 107
Fichtenkreuzschnabel 112
Fischadler 30
Fitis 88
Flussregenpfeifer 44
Flussseeschwalbe 58
Flussuferläufer 49
Gänsegeier 31
Gänsesäger 14
Gartenbaumläufer 85
Gartengrasmücke 93
Gartenrotschwanz 98
Gebirgsstelze 104
Gelbschnabel-Sturmtaucher 21
Gelbspötter 90
Gimpel 110
Girlitz 112
Goldammer 107
Goldregenpfeifer 43
Grauammer 109
Graugans 8
Grauortolan 108
Graureiher 27
Grauschnäpper 95
Großer Brachvogel 45
Großtrappe 37
Grünling 111

Grünschenkel 50
Grünspecht 70
Habicht 35
Habichtskauz 63
Häherkuckuck 60
Halbringschnäpper 97
Halsbandschnäpper 98
Haselhuhn 17
Haubenlerche 78
Haubenmeise 81
Haubentaucher 20
Hausrotschwanz 98
Haussperling 113
Heckenbraunelle 103
Heckensänger 96
Heidelerche 78
Heringsmöwe 56
Höckerschwan 9
Hohltaube 59
Isabellsteinschmätzer 100
Kaiseradler 33
Kalanderlerche 78
Kampfläufer 46
Kappenammer 108
Karmingimpel 110
Kernbeißer 110
Kiebitz 43
Kiebitzregenpfeifer 42
Klappergrasmücke 94
Kleiber 83
Kleines Sumpfhuhn 39
Kleinspecht 68
Knäkente 10
Knutt 46
Kohlmeise 82
Kolbenente 12
Kolkrabe 76
Korallenmöwe 55
Kormoran 25
Kornweihe 33
Krähenscharbe 25
Kranich 40
Krauskopfpelikan 26
Krickente 12
Kuckuck 61
Küstenseeschwalbe 58
Kuhreiher 28

Kurzfangsperber 34
Kurzzehenlerche 77
Lachmöwe 54
Lachseeschwalbe 57
Löffelente 10
Löffler 29
Mäusebussard 36
Mariskensänger 90
Marmelente 12
Mauerläufer 84
Mauersegler 65
Mehlschwalbe 80
Merlin 71
Misteldrossel 102
Mittelmeer-Steinschmätzer 100
Mittelmeer-Sturmtaucher 21
Mittelmeermöwe 56
Mittelsäger 15
Mittelspecht 68
Mönchsgeier 31
Mönchsgrasmücke 93
Moorente 13
Mornellregenpfeifer 44
Nachtigall 96
Nachtreiher 28
Nebelkrähe 76
Neuntöter 72
Odinshühnchen 49
Ohrenlerche 77
Ohrentaucher 19
Olivenspötter 90
Orpheusspötter 90
Ortolan 108
Pfeifente 11
Pfuhlschnepfe 45
Pirol 74
Prachttaucher 18
Purpurreiher 27
Rallenreiher 28
Raubseeschwalbe 57
Raubwürger 72
Rauchschwalbe 79
Raufußbussard 36
Raufußkauz 64
Rebhuhn 16
Regenbrachvogel 44
Reiherente 13
Ringdrossel 100
Ringelgans 9

Ringeltaube 59
Rötelfalke 70
Rötelschwalbe 80
Rohrammer 109
Rohrdommel 26
Rohrschwirl 92
Rohrweihe 33
Rosaflamingo 20
Rosapelikan 25
Rostgans 10
Rotdrossel 101
Rotflügel-Brachschwalbe 51
Rotfußfalke 71
Rothalstaucher 19
Rotkehlchen 96
Rotkehlpieper 105
Rotkopfwürger 73
Rotmilan 35
Rotschenkel 51
Saatkrähe 76
Säbelschnäbler 41
Samtente 13
Samtkopf-Grasmücke 95
Sanderling 47
Sandregenpfeifer 43
Schelladler 32
Schellente 14
Schilfrohrsänger 91
Schlagschwirl 92
Schlangenadler 32
Schleiereule 61
Schmarotzerraubmöwe 52
Schmutzgeier 31
Schnatterente 11
Schneeammer 106
Schneefink 114
Schreiadler 32
Schwanzmeise 83
Schwarzhalstaucher 20
Schwarzkehlchen 99
Schwarzkopfmöwe 55
Schwarzmilan 35
Schwarzschnabel-Sturmtaucher 21
Schwarzspecht 69
Schwarzstirnwürger 73
Schwarzstorch 23
Seeadler 35
Seeregenpfeifer 43
Seidenreiher 27

Seidensänger 88
Seidenschwanz 106
Sichelstrandläufer 46
Sichler 29
Silberreiher 27
Singdrossel 101
Singschwan 9
Skua 52
Sommergoldhähnchen 87
Spatelraubmöwe 52
Sperber 34
Sperbergrasmücke 94
Spießente 11
Sprosser 96
Star 102
Steinadler 33
Steinhuhn 16
Steinkauz 62
Steinrötel 98
Steinschmätzer 99
Steinsperling 114
Steinwälzer 45
Stelzenläufer 41
Steppenmöwe 56
Steppenweihe 34
Sterntaucher 18
Stieglitz 112
Stockente 11
Sturmmöwe 55
Sturmschwalbe 22
Südraubwürger 73
Sumpfläufer 46
Sumpfmeise 81
Sumpfohreule 63
Sumpfrohrsänger 91
Tafelente 12
Tannenhäher 75
Tannenmeise 80
Teichhuhn 39
Teichrohrsänger 91
Teichwasserläufer 50
Temminckstrandläufer 47
Thorshühnchen 49
Trauerente 14
Trauermeise 81
Trauerschnäpper 97
Trauerseeschwalbe 57
Triel 40
Trottellumme 53
Tüpfelsumpfhuhn 38

Türkentaube 60	Weißbart-Seeschwalbe 58	Zwergohreule 62
Turmfalke 70	Weißflügel-Seeschwalbe 57	Zwergsäger 14
Turteltaube 60	Weißkopf-Ruderente 15	Zwergscharbe 24
Uferschnepfe 45	Weißrückenspecht 69	Zwergschnäpper 97
Uferschwalbe 79	Weißstorch 23	Zwergschnepfe 48
Uhu 62	Wellenläufer 22	Zwergschwan 9
Wacholderdrossel 101	Wendehals 68	Zwergseeschwalbe 56
Wachtel 15	Wespenbussard 31	Zwergstrandläufer 47
Wachtelkönig 38	Wiedehopf 66	Zwergsumpfhuhn 39
Waldbaumläufer 85	Wiesenpieper 104	Zwergtaucher 19
Waldkauz 63	Wiesenschafstelze 103	Zwergtrappe 37
Waldlaubsänger 89	Wiesenweihe 34	
Waldohreule 63	Wintergoldhähnchen 87	
Waldsaatgans 8	Würgfalke 71	
Waldschnepfe 48	Zaunammer 107	
Waldwasserläufer 49	Zaunkönig 86	
Wanderfalke 72	Ziegenmelker 64	
Wasseramsel 86	Zilpzalp 88	
Wasserralle 38	Zippammer 108	
Weidenmeise 81	Zwergadler 32	
Weidensperling 113	Zwergdommel 26	
Weißbart-Grasmücke 94	Zwergmöwe 55	

Index of French Names

Accenteur alpin 103
Accenteur mouchet 103
Agrobate roux 96
Aigle botté 32
Aigle criard 32
Aigle impérial 33
Aigle pomarin 32
Aigle royal 33
Aigrette garzette 27
Alouette calandre 78
Alouette calandrelle 77
Alouette des champs 78
Alouette hausse-col 77
Alouette lulu 78
Autour des palombes 35
Avocette élégante 41
Balbuzard pêcheur 30
Barge rousse 45
Barge à queue noire 45
Bec-croisé des sapins 112
Bécasse des bois 48
Bécasseau cocorli 46
Bécasseau de Temminck 47
Bécasseau falcinelle 46
Bécasseau maubèche 46
Bécasseau minute 47
Bécasseau sanderling 47
Bécasseau variable 47
Bécassine des marais 48
Bécassine double 48
Bécassine sourde 48
Bergeronnette des ruisseaux 104
Bergeronnette grise 104
Bergeronnette printanière 103
Bernache cravant 9
Bihoreau gris 28
Blongios nain 26
Bondrée apivore 31
Bouscarle de Cetti 88
Bouvreuil pivoine 110
Bruant cendrillard 108
Bruant des neiges 106
Bruant des roseaux 109
Bruant fou 108

Bruant jaune 107
Bruant mélanocéphale 108
Bruant ortolan 108
Bruant proyer 109
Bruant zizi 107
Bruant à calotte blanche 107
Busard cendré 34
Busard des roseaux 33
Busard pâle 34
Busard Saint-Martin 33
Buse féroce 36
Buse pattue 36
Buse variable 36
Butor étoilé 26
Caille des blés 15
Canard chipeau 11
Canard colvert 11
Canard pilet 11
Canard siffleur 11
Canard souchet 10
Cassenoix moucheté 75
Chardonneret élégant 112
Chevalier aboyeur 50
Chevalier arlequin 50
Chevalier cul-blanc 49
Chevalier gambette 51
Chevalier guignette 49
Chevalier stagnatile 50
Chevalier sylvain 50
Chevêche d'Athéna 62
Chocard à bec jaune 75
Choucas des tours 76
Chouette de l'Oural 63
Chouette hulotte 63
Cigogne blanche 23
Cigogne noire 23
Cincle plongeur 86
Circaète Jean-le-Blanc 32
Cisticole des joncs 93
Cochevis huppé 78
Combattant varié 46
Corbeau freux 76
Cormoran huppé 25
Cormoran pymée 24
Corneille mantelée 76
Coucou geai 60

Coucou gris 61
Courlis cendré 45
Courlis corlieu 44
Crabier chevelu 28
Crave à bec rouge 75
Cygne chanteur 9
Cygne siffleur 9
Cygne tuberculé 9
Échasse blanche 41
Effraie des clochers 61
Engoulevent d'Europe 64
Épervier d'Europe 34
Épervier à pieds courts 34
Érismature à tête blanche 15
Étourneau sansonnet 102
Faisan de Colchide 16
Faucon crécerelle 70
Faucon crécerellette 70
Faucon émerillon 71
Faucon hobereau 71
Faucon kobez 71
Faucon pèlerin 72
Faucon sacre 71
Fauvette babillarde 94
Fauvette des jardins 93
Fauvette épervière 94
Fauvette grisette 95
Fauvette mélanocéphale 95
Fauvette passerinette 94
Fauvette à gros bec 94
Fauvette à tête noire 93
Flamant rose 20
Fou de Bassan 24
Foulque macroule 39
Fuligule milouin 12
Fuligule milouinan 13
Fuligule morillon 13
Fuligule nyroca 13
Gallinule poule-d'eau 39
Garrot à oeil d'or 14
Geai des chênes 74
Gélinotte des bois 17
Glaréole à collier 51
Gobemouche gris 95
Gobemouche nain 97
Gobemouche noir 97

Gobemouche à collier 98	Huîtrier pie 42	Mouette tridactyle 54
Gobemouche à demi-collier 97	Hypolaïs des oliviers 90	Nette rousse 12
	Hypolaïs ictérine 90	Niverolle alpine 114
Goéland brun 56	Hypolaïs polyglotte 90	Nyctale de Tengmalm 64
Goéland cendré 55	Hypolaïs pâle 89	Océanite cul-blanc 22
Goéland d'Audouin 55	Ibis falcinelle 29	Océanite de Wilson 22
Goéland leucophée 56	Jaseur boréal 106	Océanite tempête 22
Goéland pontique 56	Labbe parasite 52	Oedicnème criard 40
Goéland railleur 54	Labbe pomarin 52	Oie cendrée 8
Gorgebleue à miroir 97	Labbe à longue queue 53	Oie des moissons 8
Grand Corbeau 76	Lagopède alpin 17	Oie rieuse 8
Grand Cormoran 25	Linotte mélodieuse 111	Outarde canepetière 37
Grand Harle 14	Locustelle fluviatile 92	Panure à moustaches 77
Grand Labbe 52	Locustelle luscinioïde 92	Pélican blanc 25
Grand Tétras 16	Locustelle tachetée 92	Pélican frisé 26
Grand-duc d'Europe 62	Loriot d'Europe 74	Perdrix bartavelle 16
Grande Aigrette 27	Lusciniole à moustaches 90	Perdrix grise 16
Grande Outarde 37	Macreuse brune 13	Petit-duc scops 62
Grimpereau des bois 85	Macreuse noire 14	Phalarope à bec étroit 49
Grimpereau des jardins 85	Marmaronette marbrée 12	Phalarope à bec large 49
Grive draine 102	Marouette de Baillon 39	Phragmite des joncs 91
Grive litorne 101	Marouette ponctuée 38	Pic épeiche 69
Grive mauvis 101	Marouette poussin 39	Pic épeichette 68
Grive musicienne 101	Martin-pêcheur d'Europe 66	Pic mar 68
Gros-bec casse-noyaux 110	Martinet noir 65	Pic noir 69
Grue cendrée 40	Martinet pâle 65	Pic syriaque 69
Grèbe castagneux 19	Martinet à ventre blanc 65	Pic vert 70
Grèbe esclavon 19	Merle noir 101	Pic à dos blanc 69
Grèbe huppé 20	Merle à plastron 100	Pie bavarde 75
Grèbe jougris 19	Mésange bleue 82	Pie-grièche écorcheur 72
Grèbe à cou noir 20	Mésange boréale 81	Pie-grièche grise 72
Guifette leucoptère 57	Mésange charbonnière 82	Pie-grièche méridionale 73
Guifette moustac 58	Mésange huppée 81	Pie-grièche à poitrine rose 73
Guifette noire 57	Mésange lugubre 81	
Guillemot marmette 53	Mésange noire 80	Pie-grièche à tête rousse 73
Guêpier d'Europe 67	Mésange nonnette 81	Pigeon biset 59
Gypaète barbu 30	Mésange à longue queue 83	Pigeon colombin 59
Harle huppé 15		Pigeon ramier 59
Harle piette 14	Milan noir 35	Pinson des arbres 109
Héron cendré 27	Milan royal 35	Pinson du Nord 110
Héron garde-boeufs 28	Moineau domestique 113	Pipit des arbres 105
Héron pourpré 27	Moineau espagnol 113	Pipit farlouse 104
Hibou des marais 63	Moineau friquet 113	Pipit rousseline 104
Hibou moyen-duc 63	Moineau soulcie 114	Pipit spioncelle 105
Hirondelle de fenêtre 80	Monticole merle-bleu 99	Pipit à gorge rousse 105
Hirondelle de rivage 79	Monticole merle-de-roche 98	Plongeon arctique 18
Hirondelle de rochers 79		Plongeon catmarin 18
Hirondelle rousseline 80	Mouette mélanocéphale 55	Pluvier argenté 42
Hirondelle rustique 79	Mouette pygmée 55	Pluvier doré 43
Huppe fasciée 66	Mouette rieuse 54	Pluvier grand-gravelot 43

Pluvier guignard 44
Pluvier petit-gravelot 44
Pluvier à collier interrompu 43
Pouillot fitis 88
Pouillot oriental 89
Pouillot siffleur 89
Pouillot véloce 88
Puffin cendré 21
Puffin des Anglais 21
Puffin yelkouan 21
Pygargue à queue blanche 35
Rémiz penduline 82
Roitelet huppé 87
Roitelet à triple bandeau 87
Rollier d'Europe 67
Roselin cramoisi 110
Rossignol philomèle 96
Rossignol progné 96
Rougegorge familier 96
Rougequeue noir 98
Rougequeue à front blanc 98

Rousserolle effarvatte 91
Rousserolle turdoïde 91
Rousserolle verderolle 91
Râle d'eau 38
Râle des genêts 38
Sarcelle d'été 10
Sarcelle d'hiver 12
Serin cini 112
Sittelle de Neumayer 84
Sittelle torchepot 83
Sizerin cabaret 111
Sizerin flammé 111
Spatule blanche 29
Sterne arctique 58
Sterne caspienne 57
Sterne caugek 58
Sterne hansel 57
Sterne naine 56
Sterne pierregarin 58
Tadorne casarca 10
Tadorne de Belon 10
Tarier des prés 99
Tarier pâtre 99
Tarin des aulnes 112

Tétras lyre 17
Tichodrome échelette 84
Torcol fourmilier 68
Tournepierre à collier 45
Tourterelle des bois 60
Tourterelle turque 60
Traquet isabelle 100
Traquet motteux 99
Traquet oreillard 100
Troglodyte mignon 86
Vanneau huppé 43
Vautour fauve 31
Vautour moine 31
Vautour percnoptère 31
Verdier d'Europe 111

Index of Spanish Names

Abejaruco Europeo 67
Abejero Europeo 31
Abubilla 66
Acentor Alpino 103
Acentor Común 103
Agachadiza Chica 48
Agachadiza común 48
Agachadiza Real 48
Agateador Común 85
Agateador Norteño 85
Águila calzada 32
Águila Imperial Oriental 33
Águila Moteada 32
Águila Pescadora 30
Águila Pomerana 32
Águila Real 33
Aguilucho Cenizo 34
Aguilucho Lagunero Occidental 33
Aguilucho Pálido 33
Aguilucho Papialbo 34
Aguja Colinegra 45
Aguja Colipinta 45
Alcaraván Común 40
Alcatraz Atlántico 24
Alcaudón Chico 73
Alcaudón Común 73
Alcaudón de Dorso Rojo 72
Alcaudón Norteño 72
Alcaudón Real 73
Alcotán Europeo 71
Alimoche Común 31
Alondra Común 78
Alondra Cornuda 77
Alondra Totovía 78
Alzacola 96
Ampelis Europeo 106
Anade Friso 11
Ánade Rabudo 11
Ánade Real 11
Andarríos Bastardo 50
Andarríos Chico 49
Andarríos Grande 49
Ansar Campestre 8
Ansar Careto 8
Ansar Común 8
Arao Común 53

Archibebe Claro 50
Archibebe Común 51
Archibebe Fino 50
Archibebe Oscuro 50
Arrandejo Común 74
Autillo Europeo 62
Avefría Europea 43
Avetoro Común 26
Avión Común 80
Avión Roquero 79
Avión Zapador 79
Avoceta Común 41
Avutarda Euroasiática 37
Azor Común 35
Baloncito Común 82
Barnacla Carinegra 9
Bigotudo 77
Bisbita Alpino 105
Bisbita Arbóreo 105
Bisbita Campestre 104
Bisbita Gorgirrojo 105
Bisbita Pratense 104
Búho Campestre 63
Búho Chico 63
Búho Real 62
Buitre Leonado 31
Buitre Negro 31
Buitrón Común 93
Busardo Calzado 36
Busardo Moro 36
Busardo Ratonero 36
Buscarla Fluvial 92
Buscarla Pintoja 92
Buscarla Unicolor 92
Calandria Común 78
Camachuelo Carminoso 110
Camachuelo Común 110
Canastera Común 51
Cárabo Común 63
Cárabo Uralense 63
Carbonero Común 82
Carbonero Garrapinos 80
Carbonero Lúgubre 81
Carbonero Palustre 81
Carbonero Sibilino 81
Carraca Europea 67

Carricerín Común 91
Carricerín Real 90
Carricerín Tordal 91
Carricero Común 91
Carricero Políglota 91
Cascanueces Moteado 75
Cerceta Carretona 10
Cerceta común 12
Cerceta Pardilla 12
Cernícalo Patirrojo 71
Cernícalo Primilla 70
Cernícalo Vulgar 70
Charrán Ártico 58
Charrán Común 58
Charrán Patinegro 58
Charrancito Común 56
Chocha Perdiz 48
Chochín Común 86
Chorlitejo Chico 44
Chorlitejo Grande 43
Chorlitejo Patinegro 43
Chorlito Carambolo 44
Chorlito Dorado Europeo 43
Chorlito Gris 42
Chotacabras Europeo 64
Chova de Pico Amarillo 75
Chova de Pico Rojo 75
Cigüeña Blanca 23
Cigüeña Negra 23
Cigüeñuela de Alas Negras 41
Cisne Cantor 9
Cisne Chico 9
Cisne Vulgar 9
Codorniz Común 15
Cogujada Común 78
Colimbo Ártico 18
Colimbo Chico 18
Colirrojo Real 98
Colirrojo Tizón 98
Collalba Gris 99
Collalba Isabel 100
Collalba Rubia 100
Combatiente 46
Cormorán Grande 25
Cormorán Moñudo 25
Cormorán pigmeo 24

Corneja Cenicienta 76
Correlimos Común 47
Correlimos de Temminck 47
Correlimos Falcinelo 46
Correlimos Gordo 46
Correlimos Menudo 47
Correlimos Tridáctilo 47
Correlimos Zarapitín 46
Críalo Europeo 60
Cuchara Común 10
Cuco Común 61
Cuervo Común 76
Culebrera Europea 32
Curruca Capirotada 93
Curruca Carrasqueña 94
Curruca de Cabeza Negra 95
Curruca Gavilana 94
Curruca Mirlona Oriental 94
Curruca Mosquitera 93
Curruca Pequeña 94
Curruca Zarcera 95
Escribano Ceniciento 108
Escribano Cerillo 107
Escribano de Cabeza Negra 108
Escribano Hortelano 108
Escribano Montesino 108
Escribano Nival 106
Escribano Palustre 109
Escribano Pinero 107
Escribano Soteño 107
Escribano Triguero 109
Esmerejón 71
Espátula Común 29
Estornino Pinto 102
Faisán Vulgar 16
Falaropo Picofino 49
Falaropo Picogrueso 49
Flamenco Común 20
Focha Común 39
Fumarel Aliblanco 57
Fumarel cariblanco 58
Fumarel Común 57
Gallereta Común 39
Gallo-lira Común 17
Garceta Común 27
Garceta Grande 27
Garcilla Bueyera 28
Garcilla Cangrejera 28

Garza Imperial 27
Garza Real 27
Gavilán Común 34
Gavilán Griego 34
Gaviota Cabecinegra 55
Gaviota Cana 55
Gaviota de Audouin 55
Gaviota del Caspio 56
Gaviota Enana 55
Gaviota Patiamarilla 56
Gaviota Picofina 54
Gaviota Reidora 54
Gaviota Sombría 56
Gaviota Tridáctila 54
Golondrina Común 79
Golondrina Dáurica 80
Gorrión Alpino de Alas Blancas 114
Gorrión Chillón Rayado 114
Gorrión Doméstico 113
Gorrión Molinero 113
Gorrión Moruno 113
Graja Común 76
Grajilla Común 76
Grévol Común 17
Grulla Común 40
Guión de Codornices 38
Halcón Peregrino 72
Halcón Sacre 71
Herrerillo Capuchino 81
Herrerillo Común 82
Jilguero Europeo 112
Lagópodo Alpino 17
Lavandera Blanca 104
Lavandera Boyera 103
Lavandera Cascadeña 104
Lechuza Común 61
Lúgano 112
Malvasía Cabeciblanca 15
Martín Pescador Común 66
Martinete Común 28
Milano Negro 35
Milano Real 35
Mirasol Pequeño 26
Mirlo Acuático de Garganta Blanca 86
Mirlo Común 101
Mirlo de Capa Blanca 100
Mochuelo Boreal 64
Mochuelo Europeo 62

Morito Común 29
Mosquitero Común 88
Mosquitero de los Sauces 88
Mosquitero Oriental 89
Mosquitero Silbador 89
Negrón Común 14
Negrón Especulado 13
Oropéndola Dorada Europea 74
Ostrero Euroasiático 42
Págalo Grande 52
Págalo Parásito 52
Págalo Pomarino 52
Págalo Rabero 53
Pagaza Piconegra 57
Pagaza Piquirroja 57
Paíño Boreal 22
Paíño de Wilson 22
Paíño Europeo 22
Paloma Bravía 59
Paloma Torcaz 59
Paloma Zurita 59
Papamoscas Cerrojillo 97
Papamoscas Collarino 98
Papamoscas Gris 95
Papamoscas Papirrojo 97
Papamoscas Semicollarino 97
Pardela cenicienta 21
Pardela Mediterránea 21
Pardela Pichoneta 21
Pardillo Alpino 111
Pardillo Común 111
Pardillo Sizerín 111
Pato Colorado 12
Pechiazul 97
Pelícano Ceñudo 26
Pelícano Común 25
Pepitero Común 110
Perdiz Griega 16
Perdiz Pardilla 16
Petirrojo Europeo 96
Picamaderos Negro 69
Pico Dorsiblanco 69
Pico Mediano 68
Pico Menor 68
Pico Picapinos 69
Pico Sirio 69
Pigargo Europeo 35

Pinzón Real 110
Pinzón Vulgar 109
Piquitureto Común 112
Pito Real 70
Polluela Bastarda 39
Polluela Chica 39
Polluela Pintoja 38
Porrón Bastardo 13
Porrón Europeo 12
Porrón Moñudo 13
Porrón Osculado 14
Porrón Pardo 13
Quebrantahuesos 30
Rascón Europeo 38
Reyezuelo Listado 87
Reyezuelo Sencillo 87
Roquero Rojo 98
Roquero Solitario 99
Ruiseñor 96
Ruiseñor Bastardo de Cetti 88
Ruiseñor Ruso 96
Satrecito de Cola Larga 83
Serreta Chica 14

Serreta Grande 14
Serreta Mediana 15
Silbón Europeo 11
Sisón Común 37
Sita de Eurasia 83
Sita Roquera 84
Somormujo Cuellirrojo 19
Somormujo Lavanco 20
Tarabilla común 99
Tarabilla Norteña 99
Tarro Blanco 10
Tarro Canelo 10
Terrera Común 77
Torcecuello Euroasiático 68
Tórtola Europea 60
Tórtola Turca 60
Trepa Paredes 84
Urogallo Común 16
Urraca de Pico Negro 75
Vencejo Común 65
Vencejo Pálido 65
Vencejo Real 65
Verdecillo 112
Verderón Común 111

Vuelvepiedras Común 45
Zampullín Común 19
Zampullín Cuellinegro 20
Zampullín Cuellirrojo 19
Zarapito Real 45
Zarapito Trinador 44
Zarcero Común 90
Zarcero Grande 90
Zarcero Icterino 90
Zarcero Pálido 89
Zorzal Charlo 102
Zorzal Común 101
Zorzal de Alas Rojas 101
Zorzal Real 101

Index of Japanese Names

アオアシシギ 50
アオガラ 82
アオサギ 27
アカアシシギ 51
アカエリカイツブリ 19
アカエリヒレアシシギ 49
アカゲラ 69
アカツクシガモ 10
アカトビ 35
アカハシカモメ 55
アカハシハジロ 12
アカマシコ 110
アシナガウミツバメ 22
アシナガワシ 32
アジサシ 58
アトリ 110
アビ 18
アマサギ 28
アリスイ 68
イエスズメ 113
イシチドリ 40
イスカ 112
イソシギ 49
イソヒヨドリ 99
イナバヒタキ 100
イヌワシ 33
イメベニヒワ 111
イワゴジュウガラ 84
イワスズメ 114
イワヒバリ 103
ウスハイイロチュウヒ 34
ウスユキガモ 12
ウズラクイナ 38
ウソ 110
ウタツグミ 101
ウタムシクイ 90
ウルアマツバメ 65
ウミアイサ 15
ウミガラス 53
エジプトハゲワシ 31
エゾライチョウ 17
エナガ 83
エリマキシギ 46
オオアカゲラ 69
オオカワラヒワ 111
オオジュリン 109
オオソリハシシギ 45

オオタカ 35
オオトウゾクカモメ 52
オオハクチョウ 9
オオハム 18
オオバン 39
オオモズ 72
オオヨシキリ 91
オカヨシガモ 11
オガワコマドリ 97
オグロシギ 45
オジロトウネン 47
オジロビタキ 97
オジロワシ 35
オタテヤブコマドリ 96
オニアジサシ 57
オニミズナギドリ 21
オナガガモ 11
オリアブウタムシクイ 90
カイツブリ 19
カオグロサバクヒタキ 100
カオジロアカゲラ 69
カオジロオタテガモ 15
カケス 74
カササギ 75
カタジロワシ 33
カッコウ 61
カベバシリ 84
カモメ 55
カラフトワシ 32
カワアイサ 14
カワウ 25
カワセンニュウ 92
カワセミ 66
カワラバト(ドバト) 59
カンムリカイツブリ 20
カンムリガラ 81
カンムリサギ 28
カンムリヒバリ 78
キアオジ 107
キアシセグロカモメ 56
キイロウタムシクイ 90
キクイタダキ 87
キジ 16
キセキレイ 104
キタヤナギムシクイ 88
キバシガラス 75
キバシリ 85

キョウジョシギ 45
キョクアジサシ 58
キリアイ 46
キレンジャク 106
キンクロハジロ 13
キンメフクロウ 64
クイナ 38
クサシギ 49
クビワツグミ 100
クマゲラ 69
クロウタドリ 101
クロガシラムシクイ 95
クロガモ 14
クロコウテンシ 78
クロジョウビタキ 98
クロヅル 40
クロトウゾクカモメ 52
クロハゲワシ 31
クロハラアジサシ 58
クロライチョウ 17
ケアシノスリ 36
ケアブシロエリハゲワシ 31
コアオアシシギ 50
コアカゲラ 68
コアジサシ 56
コオバシギ 46
コガネスズメ 113
コガモ 12
コガラ 81
コキジバト 60
コキンメフクロウ 62
コクイナ 39
コクガン 9
コサギ 27
コシアカツバメ 80
コシギ 48
コシジロイソヒヨドリ 98
コシジロウミツバメ 22
コチドリ 44
コチョウゲンボウ 71
コハクチョウ 9
コバシチドリ 44
コビトウ 24
コブハクチョウ 9
コノドジロムシクイ 94
コモンクイナ 38
コミミズク 63

ゴイサギ 28	チョウゲンボウ 70	ヒガラ 80
ゴシキヒワ 112	ツクシガモ 10	ヒゲガラ 77
ゴジュウガラ 83	ツバメ 79	ヒゲホオジロ 108
サヨナキドリ 96	ツメナガセキレイ 103	ヒゲワシ 30
サルハマシギ 46	ツルシギ 50	ヒシクイ 8
サンカノゴイ 26	トウゾクカモメ 52	ヒドリガモ 11
サンドイッチアジサシ 58	トビ 35	ヒメアカゲラ 68
シジュウカラ 82	トラフズク 63	ヒメウミツバメ 22
シマアジ 10	ニシアカガシラチョウゲンボウ 71	ヒメオオモズ 73
シマムシクイ 94	ニシイワツバメ 80	ヒメカモメ 55
シメ 110	ニシオオノスリ 36	ヒメクイナ 39
シュバシコウ 23	ニシコウライウグイス 74	ヒメクマタカ 32
ショウドウツバメ 79	ニシコクマルガラス 76	ヒメコウテンシ 77
シラガホオジロ 107	ニシズグロカモメ 55	ヒメチョウゲンボウ 70
シラコバト 60	ニシセグロカモメ 56	ヒメハイイロチュウヒ 34
シラヒゲムシクイ 94	ニシツバメチドリ 51	ヒメノガン 37
シロエリヒタキ 98	ニシツリスガラ 82	ヒメモリバト 59
シロカツオドリ 24	ニシトウネン 47	ヒメヨシゴイ 26
シロチドリ 43	ニシハイイロペリカン 26	ヒメミズナギドリ 21
シロハラアマツバメ 65	ニシヒバリ 78	ビロアドキンクロ 13
シロハラトウゾクカモメ 53	ニシブッポウソウ 67	フクロウ 63
シロビタイジョウビタキ 98	ニシノビタキ 99	ブロンズトキ 29
スゲヨシキリ 91	ニワムシクイ 93	ヘラサギ 29
スズガモ 13	ハイイロイワシャコ 16	ベニイロフラミンゴ 20
スペインスズメ 113	ハイイロウタムシクイ 89	ベニハシガラス 75
ズアオアトリ 109	ハイイロガラス 76	ベニヒワ 111
ズアオホオジロ 108	ハイイロガン 8	ナベコウ 23
ズアカモズ 73	ハイイロチュウヒ 33	ホオジロガモ 14
ズグロチャキンチョウ 108	ハイイロヒレアシシギ 49	ホシガラス 75
ズグロムシクイ 93	ハイタカ 34	ホシハジロ 12
セアカアハヤブサ 71	ハシグロクロハラアジサシ 57	ホシムクドリ 102
セアカモズ 72	ハシグロヒタキ 99	ヌマセンニュウ 92
セイタカシギ 41	ハシビロガモ 10	ヌマヨシキリ 91
セッカ 93	ハシブトアジサシ 57	マガモ 11
セリン 112	ハシブトガラ 81	ノガン 37
ソリハシセイタカシギ 41	ハシボソカモメ 54	マガン 8
タイリクハクセキレイ 104	ハジロカイツブリ 20	マキバタヒバリ 104
タカブシギ 50	ハジロクロハラアジサシ 57	ノスリ 36
タゲリ 43	ハジロコチドリ 43	マダラカンムリカッコウ 60
タシギ 48	ハタホオジロ 109	マダラヒタキ 97
タヒバリ 105	ハマシギ 47	ノドアカホオジロ 108
タンシキバシリ 85	ハマヒバリ 77	ノドグロアオジ 107
ダイサギ 27	ハヤブサ 72	ノドジロムシクイ 95
ダイシャクシギ 45	ハンエリヒタキ 97	ノハラツグミ 101
ダイゼン 42	バルカンコガラ 81	マヒワ 112
チゴハヤブサ 71	バン 39	マンクスミズナギドリ 21
チフチャフ 88	ヒガシボネリアムシクイ 89	マミジロキクイタダキ 87
チャイロツバメ 79	ヒガシメジロムシクイ 94	マミジロノビタキ 99
チュウシャクシギ 44		マミジロヨシキリ 90
チュウヒワシ 32		ムジタヒバリ 104

ムネアカタヒバリ 105
ムネアカヒワ 111
ムナジロカワガラス 86
ムナフヒタキ 95
ムラサキサギ 27
メジロガモ 13
メンフクロウ 61
モモイロペリカン 25
モリバト 59
モリヒバリ 78
モリフクロウ 63
モリムシクイ 89
ヤチセンニュウ 92
ヤツガシラ 66
ヤドリギツグミ 102
ヤブサヨナキドリ 96
ヤマシギ 48
ユキスズメ 114
ユキホオジロ 106
ユリカモメ 54
ヨアロッパアオゲラ 70

ヨアロッパアマツバメ 65
ヨアロッパウグイス 88
ヨアロッパウズラ 15
ヨアロッパオオライチョウ 16
ヨアロッパカヤクグリ 103
ヨアロッパコマドリ 96
ヨアロッパコノハズク 62
ヨアロッパジシギ 48
ヨアロッパチュウヒ 33
ヨアロッパハチクイ 67
ヨアロッパハチクマ 31
ヨアロッパヒメウ 25
ヨアロッパビンズイ 105
ヨアロッパムナグロ 43
ヨアロッパヤマウズラ 16
ヨアロッパヨシキリ 91
ヨアロッパヨタカ 64
ライチョウ 17
レバントハイタカ 34
ワキアカツグミ 101

ワシミミズク 62
ワタリガラス 76
ミコアイサ 14
ミサゴ 30
ミソサザイ 86
ミツユビカモメ 54
ミナミオオモズ 73
ミヤコドリ 42
ミヤマガラス 76
ミユビシギ 47
ミミカイツブリ 19

Index of Chinese Names

三趾滨鹬 47
三趾鸥 54
丘鹬 48
东歌林莺 94
东柳莺 89
东雀鹰 34
中斑啄木鸟 68
中杓鹬 44
中贼鸥 52
乌灰鸫 34
乌雕 32
乌鸫 101
云石斑鸭 12
云雀 78
亚高山林莺 94
仓鸮 61
侏鸬鹚 24
冠小嘴乌鸦 76
凤头鸊鷉 20
凤头山雀 81
凤头潜鸭 13
凤头百灵 78
凤头麦鸡 43
剑鸻 43
南灰伯劳 73
半领姬鹟 97
卷羽鹈鹕 26
反嘴鹬 41
叙利亚啄木鸟 69
叽喳柳莺 88
喜鹊 75
圃鹀 108
北极燕鸥 58
北贼鸥 52
北鲣鸟 24
地中海鸥 55
地中海鸌 21
原鸽 59
夜鹭 28
大天鹅 9
大山雀 82
大斑凤头鹃 60
大斑啄木鸟 69
大杜鹃 61
大白鹭 27
大红鹳 20
大短趾百灵 77

大苇莺 91
大西洋鸌 21
大鸨 37
大麻鳽 26
太平鸟 106
姬田鸡 39
姬鹬 48
家燕 79
家麻雀 113
宽尾树莺 88
寒鸦 76
小鸊鷉 19
小乌雕 32
小嘴鸻 44
小天鹅 9
小斑啄木鸟 68
小朱顶雀 111
小滨鹬 47
小田鸡 39
小苇鳽 26
小鸥 55
小鸨 37
小黑背银鸥 56
岩鸻 84
岩燕 79
岩雷鸟 17
崖沙燕 79
崖海鸦 53
平原鹨 104
庭园林莺 93
弯嘴滨鹬 46
彩鹬 29
戴胜 66
戴菊 87
扇尾沙锥 48
文须雀 77
斑头秋沙鸭 14
斑姬鹟 97
斑尾塍鹬 45
斑尾林鸽 59
斑背潜鸭 13
斑胸田鸡 38
斑脸海番鸭 13
斑腹沙锥 48
斑鸫 95
新疆歌鸲 96
旋木雀 85

星鸦 75
普通鸬 83
普通朱雀 110
普通燕鸥 58
普通秋沙鸭 14
普通秧鸡 38
普通翠鸟 66
普通鵟 36
普通鸬鹚 25
暗山雀 81
暴风海燕 22
松鸡 16
松鸦 74
林(即鸟)伯劳 73
林岩鹨 103
林柳莺 89
林百灵 78
林鹞 105
林鹬 50
棕尾鵟 36
棕扇尾莺 93
棕薮鸲 96
橄鸫 102
横斑林莺 94
橄榄篱莺 90
欧亚夜鹰 64
欧亚攀雀 82
欧亚金鸻 43
欧亚鸽 96
欧斑鸠 60
欧歌鸫 101
欧歌鸲 96
欧洲丝雀 112
欧柳莺 88
欧石鸡 16
欧金翅雀 111
欧鸬鹚 25
欧鸽 59
歌篱莺 90
毛脚燕 80
毛脚鵟 36
水鹨 105
沙(即鸟) 100
河乌 86
河蝗莺 92
沼泽山雀 81
泽鹬 50

流苏鹬 46	白额雁 8	苍雨燕 65
海鸥 55	白骨顶 39	苍鹭 27
渡鸦 76	白鹡鸰 25	苍鹰 35
游隼 72	白鹡鸰 104	草地鹨 104
湿地苇莺 91	白鹭 27	草原百灵 78
火冠戴菊 87	白鹳 23	草原石(即鸟) 99
灰伯劳 72	猛隼 21	草原鹨 34
灰山鹑 16	猎隼 71	草绿篱莺 89
灰斑鸠 60	环颈雉 16	草鹭 27
灰林鸮 63	环颈鸻 100	蒲苇莺 91
灰瓣蹼鹬 49	环颈鸻 43	蓝喉歌鸲 97
灰白喉林莺 95	秃鹫 31	蓝头圃鹀 108
灰眉岩鹀 108	秃鼻乌鸦 76	蓝矶鸫 99
灰背隼 71	穗(即鸟) 99	蓝胸佛法僧 67
灰雁 8	紫翅椋鸟 102	蚁䴕 68
灰鸻 42	红交嘴雀 112	芦苇莺 91
灰鹡鸰 104	红喉姬鹟 97	芦鹀 109
灰鹤 40	红喉潜鸟 18	花尾榛鸡 17
煤山雀 80	红喉鹨 105	蛎鹬 42
燕隼 71	红嘴山鸦 75	褐头山雀 81
燕雀 110	红嘴巨鸥 57	西域兀赞 31
牛背鹭 28	红嘴鸥 54	西红脚隼 71
琵嘴鸭 10	红头潜鸭 12	西红角鸮 62
田鹀 101	红尾鸲 98	西鸻鹬 15
疣鼻天鹅 9	红旋旋壁雀 84	角鹩䴘 19
白兀鹫 31	红背伯劳 72	角百灵 77
白喉林莺 94	红胸秋沙鸭 15	豆雁 8
白嘴端凤头燕鸥 58	红脚鹬 51	赤嘴潜鸭 12
白头硬尾鸭 15	红腹滨鹬 46	赤胸朱顶雀 111
白头鹞 107	红腹灰雀 110	赤膀鸭 11
白头鹀 33	红隼 70	赤颈䴙䴘 19
白尾海雕 35	红颈瓣蹼鹬 49	赤颈鸭 11
白尾鹞 33	红额金翅雀 112	赤鸢 35
白斑翅雪雀 114	纵纹腹小鸮 62	赤麻鸭 10
白琵鹭 29	细嘴鸥 54	赭红尾鸲 98
白眉歌鸫 101	绿啄木鸟 70	金眶鸻 44
白眉鸭 10	绿头鸭 11	金腰燕 80
白眼潜鸭 13	绿篱莺 90	金雕 33
白翅浮鸥 57	绿翅鸭 12	金黄鹂 74
白翅黄池鹭 28	翘鼻麻鸭 10	针尾鸭 11
白肩雕 33	翻石鹬 45	银喉长尾山雀 83
白背啄木鸟 69	短尾贼鸥 52	锡嘴雀 110
白背矶鸫 98	短耳鸮 63	长尾林鸮 63
白腰叉尾海燕 22	短趾旋木雀 85	长尾贼鸥 53
白腰朱顶雀 111	短趾雕 32	长耳鸮 63
白腰杓鹬 45	石雀 114	长脚秧鸡 38
白腰草鹬 49	石鸻 40	阔嘴鹬 46
白顶(即鸟) 100	矶鹬 49	雀鹰 34
白领姬鹟 98	胡兀鹫 30	雕鸮 62
白额燕鸥 56	苍头燕雀 109	雨燕 65

雪鹑 106	黄喉蜂虎 67	黑琴鸡 17
青山雀 82	黄嘴山鸦 75	黑翅长脚鹬 41
青脚滨鹬 47	黄爪隼 70	黑胸麻雀 113
青脚鹬 50	黄腿鸥 56	黑腹滨鹬 47
靴隼雕 32	黄蹼洋海燕 22	黑雁 9
须浮鸥 58	黄道眉鹀 107	黑顶林莺 93
须苇莺 90	黄雀 112	黑颈鸊鷉 20
领岩鹨 103	黄鹀 107	黑额伯劳 73
领燕鸻 51	黄鹡鸰 103	黑鸢 35
高山雨燕 65	黍鹀 109	黑鹳 23
鬼鸮 64	黑啄木鸟 69	
鸥嘴噪鸥 57	黑喉潜鸟 18	
鸲蝗莺 92	黑喉石(即鸟) 99	
鹃头蜂鹰 31	黑头林莺 95	
鹗 30	黑头鸬 108	
鹊鸭 14	黑尾塍鹬 45	
鹤鹬 50	黑斑蝗莺 92	
鸫鹨 86	黑水鸡 39	
麻雀 113	黑浮鸥 57	
黄脚银鸥 56	黑海番鸭 14	

Index of Scientific Names

aalge, Uria 53
Acanthis 111
Accipiter 34
Acrocephalus 90
Actitis 49
acuta, Anas 11
Aegithalos 83
Aegolius 64
Aegypius 31
aeruginosus, Circus 33
Alauda 78
alba, Ardea 27
alba, Calidris 47
alba, Motacilla 104
alba, Tyto 61
albellus, Mergellus 14
albicilla, Haliaeetus 35
albicollis, Ficedula 98
albifrons, Anser 8
albifrons, Sternula 56
Alcedo 66
Alectoris 16
alexandrinus, Charadrius 43
alpestris, Eremophila 77
alpina, Calidris 47
aluco, Strix 63
Anas 11
angustirostris, Marmaronetta 12
Anser 8
anser, Anser 8
Anthus 104
apiaster, Merops 67
apivorus, Pernis 31
apricaria, Pluvialis 43
Apus 65
apus, Apus 65
aquaticus, Rallus 38
Aquila 33
arborea, Lullula 78
arctica, Gavia 18
Ardea 27
Ardeola 28
Arenaria 45
aristotelis, Phalacrocorax 25

arquata, Numenius 45
arundinaceus, Acrocephalus 91
arvensis, Alauda 78
Asio 63
ater, Periparus 80
Athene 62
atra, Fulica 39
atricapilla, Sylvia 93
atthis, Alcedo 66
audouinii, Ichthyaetus 55
auritus, Podiceps 19
avosetta, Recurvirostra 41
Aythya 12
barbatus, Gypaetus 30
bassanus, Morus 24
bernicla, Branta 9
biarmicus, Panurus 77
Bombycilla 106
bonasia, Tetrastes 17
borin, Sylvia 93
Botaurus 26
brachydactyla, Calandrella 77
brachydactyla, Certhia 85
Branta 9
brevipes, Accipiter 34
Bubo 62
bubo, Bubo 62
Bubulcus 28
Bucephala 14
Burhinus 40
Buteo 36
buteo, Buteo 36
cabaret, Acanthis 111
cachinnans, Larus 56
caeruleus, Cyanistes 82
caesia, Emberiza 108
calandra, Emberiza 109
calandra, Melanocorypha 78
Calandrella 77
Calidris 46
Calonectris 21
campestris, Anthus 104
cannabina, Linaria 111
canorus, Cuculus 61

cantillans, Sylvia 94
canus, Larus 55
canutus, Calidris 46
Caprimulgus 64
carbo, Phalacrocorax 25
Carduelis 112
carduelis, Carduelis 112
Carpodacus 110
caryocatactes, Nucifraga 75
caspia, Hydroprogne 57
caudatus, Aegithalos 83
Cecropis 80
Cercotrichas 96
Certhia 85
cervinus, Anthus 105
cetti, Cettia 88
Cettia 88
Charadrius 43
cherrug, Falco 71
Chlidonias 57
Chloris 111
chloris, Chloris 111
chloropus, Gallinula 39
Chroicocephalus 54
chrysaetos, Aquila 33
cia, Emberiza 108
Ciconia 23
ciconia, Ciconia 23
Cinclus 86
cinclus, Cinclus 86
cinerea, Ardea 27
cinerea, Motacilla 104
Circaetus 32
Circus 33
cirlus, Emberiza 107
Cisticola 93
citrinella, Emberiza 107
Clamator 60
Clanga 32
clanga, Clanga 32
clangula, Bucephala 14
clypeata, Spatula 10
Coccothraustes 110
coccothraustes, Coccothraustes 110
coelebs, Fringilla 109

colchicus, Phasianus 16
collaris, Prunella 103
collurio, Lanius 72
collybita, Phylloscopus 88
Columba 59
columbarius, Falco 71
columbianus, Cygnus 9
communis, Sylvia 95
Coracias 67
corax, Corvus 76
cornix, Corvus 76
Corvus 76
Coturnix 15
coturnix, Coturnix 15
crassirostris, Sylvia 94
crecca, Anas 12
Crex 38
crex, Crex 38
crispus, Pelecanus 26
cristata, Galerida 78
cristatus, Lophophanes 81
cristatus, Podiceps 20
Cuculus 61
curruca, Sylvia 94
curvirostra, Loxia 112
cyaneus, Circus 33
Cyanistes 82
Cygnus 9
cygnus, Cygnus 9
daurica, Cecropis 80
decaocto, Streptopelia 60
Delichon 80
Dendrocopos 68
diomedea, Calonectris 21
domesticus, Passer 113
Dryocopus 69
dubius, Charadrius 44
Egretta 27
Emberiza 107
epops, Upupa 66
Eremophila 77
Erithacus 96
erythrinus, Carpodacus 110
erythropus, Tringa 50
europaea, Sitta 83
europaeus, Caprimulgus 64
excubitor, Lanius 72
fabalis, Anser 8
falcinellus, Calidris 46
falcinellus, Plegadis 29

Falco 70
familiaris, Certhia 85
ferina, Aythya 12
ferruginea, Calidris 46
ferruginea, Tadorna 10
Ficedula 97
flammea, Acanthis 111
flammeus, Asio 63
flava, Motacilla 103
fluviatilis, Locustella 92
Fringilla 109
frugilegus, Corvus 76
Fulica 39
fulicarius, Phalaropus 49
fuligula, Aythya 13
fulvus, Gyps 31
funereus, Aegolius 64
fusca, Melanitta 13
fuscus, Larus 56
galactotes, Cercotrichas 96
Galerida 78
gallicus, Circaetus 32
Gallinago 48
gallinago, Gallinago 48
Gallinula 39
Garrulus 74
garrulus, Bombycilla 106
garrulus, Coracias 67
garzetta, Egretta 27
Gavia 18
Gelochelidon 57
genei, Chroicocephalus 54
gentilis, Accipiter 35
glandarius, Clamator 60
glandarius, Garrulus 74
Glareola 51
glareola, Tringa 50
graculus, Pyrrhocorax 75
graeca, Alectoris 16
grisegena, Podiceps 19
Grus 40
grus, Grus 40
Gypaetus 30
Gyps 31
Haematopus 42
Haliaeetus 35
haliaetus, Pandion 30
heliaca, Aquila 33
hiaticula, Charadrius 43
Hieraaetus 32

Himantopus 41
himantopus, Himantopus 41
Hippolais 90
Hirundo 79
hirundo, Sterna 58
hispanica, Oenanthe 100
hispaniolensis, Passer 113
hortulana, Emberiza 108
hybrida, Chlidonias 58
Hydrobates 22
Hydrocoloeus 55
Hydroprogne 57
hypoleuca, Ficedula 97
hypoleucos, Actitis 49
ibis, Bubulcus 28
Ichthyaetus 55
icterina, Hippolais 90
Iduna 89
ignicapilla, Regulus 87
iliacus, Turdus 101
interpres, Arenaria 45
isabellina, Oenanthe 100
Ixobrychus 26
juncidis, Cisticola 93
Jynx 68
Lagopus 17
lagopus, Buteo 36
Lanius 72
lapponica, Limosa 45
Larus 55
leucocephala, Oxyura 15
leucocephalos, Emberiza 107
leucopterus, Chlidonias 57
leucorhoa, Oceanodroma 22
leucorodia, Platalea 29
leucotos, Dendrocopos 69
Limosa 45
limosa, Limosa 45
Linaria 111
livia, Columba 59
lobatus, Phalaropus 49
Locustella 92
longicaudus, Stercorarius 53
Lophophanes 81
Loxia 112
lugubris, Poecile 81

Lullula 78
Luscinia 96
luscinia, Luscinia 96
luscinioides, Locustella 92
Lymnocryptes 48
macrourus, Circus 34
major, Dendrocopos 69
major, Parus 82
Mareca 11
marila, Aythya 13
Marmaronetta 12
martius, Dryocopus 69
media, Gallinago 48
medius, Dendrocopos 68
megarhynchos, Luscinia 96
Melanitta 13
melanocephala, Emberiza 108
melanocephala, Sylvia 95
melanocephalus, Ichthyaetus 55
Melanocorypha 78
melanopogon, Acrocephalus 90
melba, Apus 65
merganser, Mergus 14
Mergellus 14
Mergus 14
meridionalis, Lanius 73
Merops 67
merula, Turdus 101
michahellis, Larus 56
Microcarbo 24
migrans, Milvus 35
Milvus 35
milvus, Milvus 35
minimus, Lymnocryptes 48
minor, Dendrocopos 68
minor, Lanius 73
minuta, Calidris 47
minutus, Hydrocoloeus 55
minutus, Ixobrychus 26
modularis, Prunella 103
monachus, Aegypius 31
monedula, Corvus 76
montanus, Passer 113
montanus, Poecile 81
Monticola 98
Montifringilla 114
montifringilla, Fringilla 110

morinellus, Charadrius 44
Morus 24
Motacilla 103
muraria, Tichodroma 84
Muscicapa 95
muta, Lagopus 17
naevia, Locustella 92
naumanni, Falco 70
nebularia, Tringa 50
Neophron 31
Netta 12
neumayer, Sitta 84
niger, Chlidonias 57
nigra, Ciconia 23
nigra, Melanitta 14
nigricollis, Podiceps 20
nilotica, Gelochelidon 57
nisoria, Sylvia 94
nisus, Accipiter 34
nivalis, Montifringilla 114
nivalis, Plectrophenax 106
noctua, Athene 62
Nucifraga 75
Numenius 44
Nycticorax 28
nycticorax, Nycticorax 28
nyroca, Aythya 13
oceanicus, Oceanites 22
Oceanites 22
Oceanodroma 22
ochropus, Tringa 49
ochruros, Phoenicurus 98
oedicnemus, Burhinus 40
Oenanthe 99
oenanthe, Oenanthe 99
oenas, Columba 59
olivetorum, Hippolais 90
olor, Cygnus 9
onocrotalus, Pelecanus 25
orientalis, Phylloscopus 89
Oriolus 74
oriolus, Oriolus 74
ostralegus, Haematopus 42
Otis 37
Otus 62
otus, Asio 63
Oxyura 15
pallida, Iduna 89
pallidus, Apus 65
palumbus, Columba 59

palustris, Acrocephalus 91
palustris, Poecile 81
Pandion 30
Panurus 77
paradisaea, Sterna 58
parasiticus, Stercorarius 52
Parus 82
parva, Ficedula 97
parva, Zapornia 39
Passer 113
pelagicus, Hydrobates 22
Pelecanus 25
pendulinus, Remiz 82
penelope, Mareca 11
pennatus, Hieraaetus 32
percnopterus, Neophron 31
Perdix 16
perdix, Perdix 16
peregrinus, Falco 72
Periparus 80
Pernis 31
Petronia 114
petronia, Petronia 114
phaeopus, Numenius 44
Phalacrocorax 25
Phalaropus 49
Phasianus 16
philomelos, Turdus 101
Phoenicopterus 20
Phoenicurus 98
phoenicurus, Phoenicurus 98
Phylloscopus 88
Pica 75
pica, Pica 75
Picus 70
pilaris, Turdus 101
Platalea 29
platyrhynchos, Anas 11
Plectrophenax 106
Plegadis 29
Pluvialis 42
Podiceps 19
Poecile 81
polyglotta, Hippolais 90
pomarina, Clanga 32
pomarinus, Stercorarius 52
Porzana 38
porzana, Porzana 38
pratensis, Anthus 104

pratincola, Glareola 51
Prunella 103
Ptyonoprogne 79
Puffinus 21
puffinus, Puffinus 21
pugnax, Calidris 46
purpurea, Ardea 27
pusilla, Zapornia 39
pygargus, Circus 34
pygmeus, Microcarbo 24
Pyrrhocorax 75
pyrrhocorax, Pyrrhocorax 75
Pyrrhula 110
pyrrhula, Pyrrhula 110
querquedula, Spatula 10
ralloides, Ardeola 28
Rallus 38
Recurvirostra 41
Regulus 87
regulus, Regulus 87
Remiz 82
ridibundus, Chroicocephalus 54
Riparia 79
riparia, Riparia 79
Rissa 54
roseus, Phoenicopterus 20
rubecula, Erithacus 96
rubetra, Saxicola 99
rubicola, Saxicola 99
ruficollis, Tachybaptus 19
rufina, Netta 12
rufinus, Buteo 36
rupestris, Ptyonoprogne 79
rustica, Hirundo 79
rusticola, Scolopax 48
sandvicensis, Thalasseus 58

saxatilis, Monticola 98
Saxicola 99
schoeniclus, Emberiza 109
schoenobaenus, Acrocephalus 91
scirpaceus, Acrocephalus 91
Scolopax 48
scops, Otus 62
semitorquata, Ficedula 97
senator, Lanius 73
Serinus 112
serinus, Serinus 112
serrator, Mergus 15
sibilatrix, Phylloscopus 89
Sitta 83
skua, Stercorarius 52
solitarius, Monticola 99
Spatula 10
spinoletta, Anthus 105
Spinus 112
spinus, Spinus 112
squatarola, Pluvialis 42
stagnatilis, Tringa 50
stellaris, Botaurus 26
stellata, Gavia 18
Stercorarius 52
Sterna 58
Sternula 56
strepera, Mareca 11
Streptopelia 60
striata, Muscicapa 95
Strix 63
Sturnus 102
subbuteo, Falco 71
svecica, Luscinia 97
Sylvia 93
syriacus, Dendrocopos 69

Tachybaptus 19
Tadorna 10
tadorna, Tadorna 10
tarda, Otis 37
temminckii, Calidris 47
Tetrao 16
Tetrastes 17
Tetrax 37
tetrax, Tetrax 37
tetrix, Tetrao 17
Thalasseus 58
Tichodroma 84
tinnunculus, Falco 70
torquatus, Turdus 100
torquilla, Jynx 68
totanus, Tringa 51
tridactyla, Rissa 54
Tringa 49
trivialis, Anthus 105
trochilus, Phylloscopus 88
Troglodytes 86
troglodytes, Troglodytes 86
Turdus 100
turtur, Streptopelia 60
Tyto 61
Upupa 66
uralensis, Strix 63
urbicum, Delichon 80
Uria 53
urogallus, Tetrao 16
Vanellus 43
vanellus, Vanellus 43
vespertinus, Falco 71
viridis, Picus 70
viscivorus, Turdus 102
vulgaris, Sturnus 102
yelkouan, Puffinus 21
Zapornia 39

Additional Copyright Terms

Please note the following copyright terms applying to the individual fotos, drawings, and grafics contained in this e-book:

As far as known, the names of the authors (photographers, graphic artists and other creators) of the individual photos, drawings, graphics, maps and other works are placed directly next to the respective images (Attribution).

Moreover, every photo, drawing, graphic, map and each other work is labeled with an abbreviation that refers to the license under which the work has been reproduced here. These abbreviations mean:

AU	AVITOPIA holds the copyright. All rights reserved.
LIC	AVITOPIA has obtained certain licenses from the holder of the rights of use. All rights reserved.
S2.0	To the best of the knowledge of AVITOPIA this image is subject to the Creative Commons Attribution Share Alike License 2.0 the complete conditions of which can be found at Creativecommons.org/licenses/by-sa/2.0/deed.en The author has some rights reserved.
S2.5	To the best of the knowledge of AVITOPIA this image is subject to the Creative Commons Attribution Share Alike License 2.5 the complete conditions of which can be found at Creativecommons.org/licenses/by-sa/2.5/deed.en The author has some rights reserved.
S3.0	To the best of the knowledge of AVITOPIA this image is subject to the Creative Commons Attribution Share Alike License 3.0 the complete conditions of which can be found at Creativecommons.org/licenses/by-sa/3.0/deed.en The author has some rights reserved.
S4.0	To the best of the knowledge of AVITOPIA this image is subject to the Creative Commons Attribution Share Alike License 4.0 the complete conditions of which can be found at Creativecommons.org/licenses/by-sa/4.0/deed.en The author has some rights reserved.
A2.0	To the best of the knowledge of AVITOPIA this image is subject to the Creative Commons Attribution License 2.0 the complete conditions of which can be found at Creativecommons.org/licenses/by/2.0/deed.en The author has some rights reserved.
A3.0	To the best of the knowledge of AVITOPIA this image is subject to the Creative Commons Attribution License 3.0 the complete conditions of which can be found at Creativecommons.org/licenses/by/3.0/deed.en The author has some rights reserved.
PD	To the best of the knowledge of AVITOPIA this image is in the public domain. However, AVITOPIA does not accept any liability if you use such an image. It is the obligation of the user, to verify the absence of any right reservations under his/her jurisdiction.

CPSIA information can be obtained
at www.ICGtesting.com
Printed in the USA
BVHW092056080621
609008BV00003B/449